Between MITI and the Market

Between
MITI and the Market

JAPANESE INDUSTRIAL POLICY
FOR HIGH TECHNOLOGY

Daniel I. Okimoto

STANFORD UNIVERSITY PRESS
Stanford, California

Studies in International Policy
Lawrence J. Lau and Daniel I. Okimoto, General Editors
Sponsored by the Northeast Asia–United States Forum on International Policy
of **Stanford University**

Stanford University Press, Stanford, California
© 1989 by the Board of Trustees of the Leland Stanford Junior University
Printed in the United States of America
Original printing 1989
Last figure below indicates year of this printing:
99   98   97   96   95   94   93   92   91

CIP data appear at the end of the book

FOR NANCY

Acknowledgments

I would like to acknowledge, with gratitude, the support and assistance of many individuals and institutions. The Japan–United States Friendship Commission and the United States–Japan Foundation provided financial support for the research project from which this book emerged; a special word of thanks to Dr. Robert E. Ward, former chairman of the Friendship Commission, for all his help over the years.

Stanford University's Northeast Asia–United States Forum on International Policy, under whose auspices I wrote this manuscript, also provided summer support, research assistance, computer facilities, photocopying services, and a very stimulating environment. Of the many colleagues and friends at the forum, I should like to mention the help of Gerry Bowman, Yvonne Brown, Kris England, Stacey Green, Rosemary Hamerton-Kelley, Josephine Harris, John W. Lewis, Helen Morales, and Thomas P. Rohlen. How lucky I am to be surrounded by such competent and delightful people! Thanks also to John Rogers, Kyle Koehler, and Lili Shaver for their invaluable research assistance, and to Grant Barnes, Muriel Bell, John Feneron, and John Ziemer at Stanford University Press.

Several colleagues took the time to read the entire manuscript in earlier versions and give me the benefit of their insights: Hugh Patrick, Gary Saxonhouse, Ezra Vogel, and Kōzō Yamamura. I would also like to thank all the MITI officials, who patiently answered my many questions and gave generously of their time and thoughts, especially Takeshi Isayama, Fumihiko Kato, and Norihiko Ishiguro. Of course, I alone am accountable for any factual errors or misinterpretations.

A final and special expression of thanks to my family for respecting the quiet of many work days sequestered in my study and for putting up with my prolonged preoccupation with this manuscript. I owe a particular debt to my wife, Nancy, without whose help I could not have completed this book. Not only did she give me encouragement to continue in moments of frustration; she also came to my rescue when I needed expert assistance on the word processor. It is fitting that I dedicate this book to her, with love and gratitude.

D.I.O.

Contents

Tables and Figures

Preface

Over the postwar period, the scope of industrial policy has expanded markedly. Governments in virtually all advanced industrial countries have extended the visible hand of the state in assisting specific industries or individual companies with depreciation allowances for new plant facilities, subsidies for research and development, import protection against foreign-made products, public works and procurements, unemployment payments, and mediation in labor-management conflicts. As in the 1930's, the state has become entangled in matters that proponents of market capitalism like Adam Smith believed should be left to the decentralized decisions of individuals and corporations.

Several reasons can be cited for the worldwide expansion of industrial policy. Perhaps the most fundamental has been the tremendous proliferation of ties of economic interdependence beyond all levels previously known in world history. As a result of the ever-widening web of economic interdependence, individual nation-states have had to cope with the headaches of adjusting to the changing international division of labor. Governments everywhere have had to respond to political cries for help in maintaining acceptable levels of employment, generating jobs in depressed regions, administering labor retraining programs, and slowing down the speed of structural change. The task has not been made easier by the worldwide slowdown in rates of economic growth, precipitated in the mid-1970's by the OPEC oil embargo. Slower growth rates have exacerbated socioeconomic problems at home, intensifying the political pressures on national governments to expand the scope of market intervention.

Unfortunately, the results of greater government involvement have been mixed. Although it has helped certain countries cope with the dislocations brought about by slower growth rates, industrial policy has also caused or exacerbated a number of other problems, including distortions in the allocation of capital and labor and severe trade conflicts that threaten to destroy the postwar system of free trade. Of the industrial countries that have relied heavily on industrial policy, it is hard to find one that has not had to pay the high price of economic inefficiency or trade tensions.

Only Japan is cited as an unambiguous success story. In Japan alone have the benefits of industrial policy been thought clearly to outweigh the costs. The effectiveness of industrial policy is revealed in the successful emergence of one industry after another as world-class competitors: for example, steel, automobiles, and semiconductors. Foreign countries fear that a number of still-developing industries—like biotechnology, telecommunications, and information processing—will enjoy the same success. The reason Japan strikes fear in the hearts of foreign producers, even in areas of high technology, is the alleged effectiveness of government targeting.

But is industrial policy the main reason for Japan's competitiveness? Can it be given credit for Japan's economic achievements? What about the successes of the non-targeted industries, such as precision equipment and consumer electronics? And why is so little mention made of the problems encountered in such targeted sectors as space, large jet aircraft, and petrochemicals? The successes of non-targeted industries and the problems in certain targeted sectors suggest that the reasons for Japan's spectacular track record go well beyond the realm of industrial policy—into broad areas of the political economy as a whole. What are those areas? And what accounts for the relative effectiveness of Japanese industrial policy, compared to other industrial states? What makes Japanese industrial policy different?

In the analysis that follows, I shall try to identify the reasons for the comparative effectiveness of Japanese industrial policy in high technology. I shall answer the following questions: What attitude do Japanese leaders take toward state intervention in the marketplace? What, concretely, is the Ministry of International Trade and Industry (MITI) doing to promote the development of high technology? How has the organization of the private sector contributed to MITI's capacity to intervene effectively without getting permanently entangled and without causing the market to malfunction? Since industrial policy is the product largely of responses to political demands, what is it about Japan's political system that explains the insulation of industrial policy from the pulling and hauling of interest-group politics?

In the chapters that follow, an attempt will be made to come up with satisfactory answers to these and other questions. The book's approach and line of argument diverge from those taken by frequently cited works such as Eugene Kaplan's classic articulation of the concept of "Japan, Incorporated" in *Japan: The Government-Business Relationship* and Chalmers Johnson's concept of the developmental state in *MITI and the Japanese Miracle: The Growth of Industrial Policy, 1925–1975.* It places much less emphasis on MITI's vision and power as the central explanation for the effectiveness of Japanese industrial policy. Instead of focusing on MITI itself, it tries to cast MITI in the broader context of the complex system within which it operates. It includes an analysis of MITI's relationship to the Liberal-Democratic Party (LDP) and to the private sector, the relevance of certain characteristics of Japanese industrial organization, distinctive sociocultural factors, and the LDP-dominated political system. By broadening the focus of analysis, this book seeks not only to provide a nuanced picture of the Japanese political economy in all its complexity but also to convey some sense of its similarities to and differences from those of Western industrial countries. In many respects, Japanese capitalism is different from the capitalism that developed in the West, and some of those differences can be understood by examining Japanese industrial policy.

Between MITI and the Market

ONE

Between the Market and the State:
The Rationale

The term "Japan, Incorporated," often used to characterize Japan's political economy, conjures up images of a big and ubiquitous state, instinctively distrustful of laissez-faire capitalism and reliant on centralized planning and administrative guidance for control.[1] The scope of state intervention is exceedingly wide, covering areas considered in the United States to be best left to the "invisible hand" of the marketplace. The Ministry of International Trade and Industry (MITI), the so-called central headquarters for "Japan, Inc.," is said to favor vigorous intervention over a laissez-faire approach.[2]

If this characterization is accurate, the heavy-handedness of the Japanese state comes perilously close to behavior more commonly ascribed to centrally planned economies, which have not been noted for economic dynamism or efficiency. According to Adam Smith, excessive state intervention is bound to disrupt the finely tuned mechanism of self-regulation built into market economies, including especially the capacity to allocate labor and capital efficiently.

Is the Japanese state as disdainful of the market mechanism as the stereotype of "Japan, Inc." conveys? Does the state intervene as frequently and indiscriminately as the stereotype implies? On what conceptual or philosophical grounds are Japanese industrial policies based? In what ways is the rationale similar to, or different from, that of the U.S. government?

It is clear that MITI bureaucrats do not regard the market as sacrosanct.[3] They feel fewer ideological inhibitions about trespassing on the preserve of the private sector than their U.S. counterparts. Although capitalism is considered the best economic system yet devised, its imperfections are clearly understood. MITI officials realize that the market mechanism cannot be expected to generate economic outcomes that are always in the nation's best interests. To further the collective good, unfettered market forces need to be harnessed and guided by the visible hand of the state.

In spite of its skepticism about the reliability of Adam Smith's invisible hand, the Japanese state can be considered among the world's smallest—a "minimalist" state, as it were—in terms of the size of the "bite" it takes out of the economy. Consider such standard, unobtrusive indicators as tax revenues and fiscal expenditures. Japan's tax rate is less burdensome, at 27 percent of gross domestic product (GDP), than that of the United States, at 29 percent (prior to the 1987 tax reforms), and much less onerous than West Germany's at 38 percent, Great Britain's at 39 percent, Italy's at 41 percent, and France's at 48 percent.[4] Of 23 member nations in the Organization for Economic Cooperation and Development (OECD), Japan ranked 21st in terms of the ratio of tax revenues to GDP, with only Spain and Turkey lower. And if government revenues are smaller in Japan, so too are fiscal expenditures:[5] 16.8 percent of gross national product (GNP) in Japan, compared to 23.9 percent in America, 31.5 percent in the United Kingdom (U.K.), and 40.2 percent in Italy.[6]

Moreover, state ownership in key industries like steel, shipbuilding, railways, automobiles, aircraft, airlines, electronics, telecommunications, and banking is far smaller in Japan than in France, Italy, or the United Kingdom.[7] Instead of nationalizing industries (especially those in trouble), the Japanese government has moved to "privatize" the comparatively few that it once owned, like the deficit-ridden national railways. It has even auctioned off most of its stock in Japan Airlines and Nippon Telegraph and Telephone (NTT), a healthy public corporation in the telecommunications sector. As in the case of the United States under Ronald Reagan and England under Margaret Thatcher, Japan under Yasuhiro Nakasone made the decision to scale back state ownership in the belief that exposure to the discipline of market competition would force public enterprises to become more efficient. Contrary to widely held stereotypes, therefore, the Japanese state is actually smaller in terms of revenues, outlays, and equity ownership than most capitalist countries in the West, which are thought to be rooted more firmly in the market.

Consider a different indicator of government intrusiveness: namely, regulatory control. In certain areas of regulation, the arm of the Japanese government is neither as long nor as coercive as that of the U.S. federal government.[8] Look at interstate commerce, occupational safety and health, affirmative action and other social programs, export controls over dual-purpose (military and civilian) technology, espionage and sedition, national security, defense contracting, and anti-trust: over all these domains the regulatory hand of the state is far heavier in the United States. Here again, as in the areas of taxation, fiscal expenditures, and state ownership, the Japanese government might be described as "minimalist," not "dirigiste."

The concept of a "minimalist government" contradicts the more prevalent image of a big and ubiquitous state associated with "Japan, Incorporated." Which is the "real" Japan? The minimalist government that is restrained in its power and reliant on the private sector's compliance? Or the dirigiste state that interprets its public mandate broadly and takes action whenever it feels the market must be subordinated to its will? To answer this question requires that we analyze the underlying rationale of Japanese industrial policy, especially the view of the market taken by higher civil servants within MITI.

Industrial Policy: Inconsistencies and Costs

The Myth of Coherence

Most studies of Japanese industrial policy convey the impression that it is coherent, effective, and far-sighted.[9] The problem with these studies is that they tend to focus only on Japan's success stories—steel, automobiles, electrical power generation, and shipbuilding—and not on the less successful sectors—petrochemicals, wholesale and retail distribution, aerospace, and large commercial jet aircraft. Occasionally, a few examples of failure are cited; but even when they are, one gets the impression that the analysts have searched long and hard to ferret out the aberrant cases and that the exceptions merely underscore the general rule.[10]

The level of coherence ascribed to Japanese industrial policy tends to be overstated. Not enough effort is made to differentiate policies across industrial sectors; what holds true for the steel industry may not for software. Nor is sufficient attention paid to variations within sectors; in high technology, the policies designed for opto-electronics differ from those for space and aircraft. Even in a specific industry, like semiconductors, industrial policies may change over time and across individual companies. Contrary to conventional wisdom, therefore, Japanese industrial policy may lack the consistency and vision often attributed to it.

The perception of coherence may be attributable, in part, to the conspicuous absence of coherence in the industrial policies of other countries. Compared to the industrial policies of Italy and Great Britain, Japan's seems almost like a model of clarity and technical rationality.[11] The contrast is especially striking if the scope of industrial policy is confined to the manufacturing sectors under MITI's jurisdiction. But clarity, consistency, and effectiveness are relative terms, and if Italy, Great Britain, or even France is the point of reference, the liberal use of such terms may mean relatively little. They may reveal less about the conceptual tidiness of industrial policy in Japan than they do about its ineptly conceived and disorganized nature in other countries.

On close scrutiny, industrial policy in Japan is less consistent than is commonly believed, especially if sectors outside MITI's scope of authority are included. Japanese industrial policy consists of a motley assortment of policy measures, pragmatically devised to address the diverse, ever-changing, and sometimes conflicting needs of a broad range of industries at different points in time. As might be expected, this collage of industry-specific policies encompasses elements that may be incompatible and even contradictory in nature.

By its very nature, industrial policy everywhere tends to be an agglomeration of ad hoc responses to the special circumstances and needs of miscellaneous sectors. The various industries to whose demands the government is forced to respond seldom have a common set of interests and goals, much less a unified set of policy instruments. Under such circumstances, harmonious integration is almost impossible to achieve.

It is especially elusive when the policy-making process is besieged by a multitude of special-interest groups clamoring for public policies favorable to their narrow self-interest. Aggregating the cacophonous demands virtually guarantees that some measure of coherence will be lost.[12] Considering also the complexity of bureaucratic and organizational behavior—the many agencies, bureaus, and divisions that have to be consulted, the infighting that inevitably takes place, the rigid rules and procedures that have to be followed—the emergence of policy inconsistencies is not at all surprising.

Errors and Costs

Most studies of Japanese industrial policy leave the impression that it has been nearly error-free and costless. Nothing could be further from the truth. One need not look far to find pockets of conspicuous inefficiency in the Japanese economy, many of which are either directly attributable to, or exacerbated by, Japanese industrial policy. Agriculture, livestock, and runaway land prices are probably the best known, but others include sugar refining, confectionaries, food processing, coal, lumber processing, silkworm cultivation, tobacco, aluminium refining, space and aeronautics, defense production, health services, retail distribution, and segments of the financial services.

How is it that Japanese industrial policy can combine sectors of extraordinary efficiency and inefficiency? The answer to this conundrum is presented fully in Chapter 4; it is to be found in the nature of the relationship between the Liberal Democratic Party (LDP), the bureaucracy, and industry. The problem in all of the above cases is that parochial interests have managed to gain a significant voice through LDP Diet members in industrial policies of direct relevance to their vested interests.[13] In most cases, such pockets of inefficiency lie outside MITI's jurisdiction, falling into the domain of such ministries as Agriculture, Fisheries, and Forestry, Construction, Health and Welfare, Transportation, Finance, and the Japan Defense Agency.

Of course, one can also find striking examples of economic inefficiency even under the umbrella of MITI's industrial policy (though, as pointed out already, MITI is far more successful at protecting public policy from being dominated by parochial politics than other ministries). The coal industry is but one example. Owing to the coal industry's political influence in certain regions like Hokkaido, Kyushu, and Ibaraki Prefecture, MITI has been locked for a long time into providing large subsidies. The subsidies cannot be justified on economic grounds, but politics prevents MITI from abolishing them. Similarly, MITI provides handsome subsidies to Japan's textile industry. Who winds up paying for such economically unjustified industrial policy? The taxpayer.

The myth of MITI omniscience is also belied by costly mistakes that have been made. Take the petrochemical industry. In the mid-1960's, MITI's ambitious plans to create a world-class petrochemical industry seemed like a sound and far-sighted idea. The petrochemical industry looked like a sure winner, a high-growth industry capable of generating high value added and certain to move Japan's industrial structure to the next logical phase of international comparative advantage. If Japanese companies invested heavily in new plant facilities, they would be in a position to capture their share of the burgeoning world market. MITI facilitated the industry's rapid development by providing, among other things, favorable incentives for heavy capital investments.

MITI's vision, however, failed to materialize. The first oil crisis hit not long after a major expansion of new plant capacity. As the costs of energy spiraled, Japan's petrochemical industry—utterly dependent on imported oil—found itself hard-pressed to compete with foreign manufacturers. The recession into which the world was plunged left Japan's petrochemical industry with large excess capacity. In hindsight, the rush to put a petrochemical complex in place could be considered a matter of bad timing, or perhaps less charitably, poor planning.

Here was a case where Japanese industrial policy led directly to serious structural problems, which required offsetting policy measures—applied over a sustained period—to correct. Even after making structural adjustments, the petrochemical industry's long-term capacity to compete with oil-rich foreign producers is in doubt. Not that a decentralized market system would have averted this error; the OPEC oil embargo caught everyone by surprise. But Japanese industrial policy exacerbated the problem by pushing the installation of new plant facilities ahead at a faster pace than would have been sustained under a laissez-faire system.

The politics of industrial policy are such that it can easily lead to deepening government involvement in the economy, as larger clusters of vested interests come to depend on government support in one form or another. Once protection or assistance is extended to one group, others are bound to ask for similar favors. This "ratcheting upward" is common in countries like the United States, where old-line industries are experiencing problems and where the institutional mechanisms for effective interest aggregation are not strong enough to contain "me too" lobbying.

Certain types of industrial policy measures appear to trigger "me too" responses: outright grants or subsidies, protection against foreign imports, countercyclical stabilization measures, discriminatory tax treatment, preferential financing, regional development, and public procurements. Once a government good or service of this kind is given to one industry, a precedent is set, and other beleaguered industries come rushing to put in their claims for similar treatment. This is especially true in the United States and parts of Western Europe.

About the only area in which the contagion effect is relatively contained is the area of national security, a special category of industrial policy. But even here, the boundaries between commercial and military concerns have begun to blur, as the relationship between commercial competitiveness in high technology, the capacity to innovate, and state-of-the-art military weapons systems has grown increasingly interdependent and complex. The U.S. semiconductor industry has sought government subsidies on national security grounds, citing the importance of a viable and innovative commercial semiconductor industry to the country's capacity to maintain the world's military balance of power.[14] If it receives support on such grounds, other high-technology industries deemed of similar strategic importance, such as supercomputers, may seek to follow suit. Such developments, if they take place, could have an enormous impact on the nature of industrial policy.

Unintended Consequences

Industrial policies often produce unintended and perverse consequences that must be offset by a package of countermeasures. Industrial policy measures that are especially apt to create distortions include sectoral targeting, investment guidelines, industrial restructuring, regulatory controls, and government-sanctioned cartels, all of which MITI has utilized at one time or another.[15]

MITI has had to organize temporary antirecession cartels for certain industries, ostensibly to control excessive competition—or, in other words, to correct for the aftereffects of excessive capital investment. The formation of temporary antirecession cartels is permitted under the Temporary Measures Law for the Stabilization of Specific Depressed Industries and the Temporary Measures Law for the Structural Adjustment of Specific Industries. The rationale for antirecession cartelization is that it preempts fratricidal warfare; it keeps the level of market concentration from increasing; and the bankruptcy of big corporations would have serious ripple effects for many small to medium-sized subcontractors, which depend on continuing orders from parent corporations. In the long run, MITI officials used to argue, the imposition of some control over excessive competition through the formation of temporary antirecession cartels was necessary to ensure that healthy competition would be sustained.

If cartels are needed to cope with excessive competition generated in part by Japanese industrial policy, they also give rise to problems that require further policy correction. Cartels have an insidious way of begetting more cartels, as Kōzō Yamamura points out.[16] The number of legal cartels rose from 162 in 1955 to around 1,000 in the late 1960's and early 1970's. The figures have fallen sharply since then, but MITI continues to rely on them to stabilize business conditions for certain industries, particularly small and medium-sized enterprises. When antirecession cartels are used in conjunction with import restrictions, they give rise to charges of unfair trade practices made by foreign companies trying to enter the Japanese market. This, in turn, usually pulls the government further into the role of mediating trade conflicts. A vicious circle of deepening government involvement is thus triggered by industrial policy.

Ryūtarō Komiya calls attention to another case of internal inconsistency. To promote exports and thus ease Japan's chronic balance of payments deficit, the government used to bestow special tax breaks (until 1963) for income earned from exports. Viewed in isolation, the tax credit made sense; but the problem was that export incentives worked against Japan's acceptance of voluntary export restraints (VERs). The presumed benefits of export-related tax privileges were negated by the existence of export ceilings, and the upward shift of the export supply curve only served to increase the special premium accruing to those operating under export quotas, with little or no stimulus for export expansion or relief for balance of payments deficits.[17] What industrial policy sought to accomplish with one hand was canceled out by industrial policy measures administered by the other.

Industrial Policy: A Definition

To discuss its conceptual underpinnings, we must first define industrial policy. Too often the term *industrial policy* is carelessly used without a clear and rigorous explication of its meaning. Sometimes it is used interchangeably with the concept of centralized economic planning, whether it be indicative planning for aggregate growth or sectoral planning for targeted industries.[18] Sometimes it is associated with government attempts to reshape industrial structure and allocate resources among various sectors.[19] Other times it takes the form of supply-side measures, especially tax policies, which are distinct from the macroeconomics of demand-side management.[20]

The definition adopted here involves the government's use of its authority and resources to administer policies that address the needs of specific sectors and industries (and, if necessary, those of individual companies) with the aim of raising the productivity of factor inputs.[21] By contrast, public policies dealing with the economy as a whole, not just with its microindustrial parts, fall into the domain of macroeconomics. Between the two poles—industrial policy and macroeconomics—are gray areas that can be grouped at either end of the continuum, depending on the uses to which they are put. Fiscal budgets, for example, have an impact on the whole economy, but individual items in the budget—like public procurements or research and development subsidies—can be used to promote specific industries and hence ought to be included in the category of industrial policy. A corporate flat tax applied equally across all industrial sectors might be considered a macroeconomic measure, even though its impact undoubtedly varies by sector. Corporate taxes that levy widely disparate rates on industries, on the other hand, can be regarded as one of the most widely used instruments of industrial policy.

The defining characteristics of industrial policy, then, is the custom design of policy instruments to fit the differing priorities, needs, and circumstances of individual industries, particularly with respect to factor inputs. It is based on the premise that certain outcomes cannot be achieved by relying solely on the invisible hand of Adam Smith's market, or on aggregate, macroeconomic measures. In complex industrial economies featuring a high degree of sectoral differentiation and international interdependence, virtually every country has adopted its own set of industrial policies as defined above.

Industrial policy is utilized to achieve a variety of national goals. For the United States, the highest-priority goal (setting aside national security) may be the creation or maintenance of full employment. In Japan, a single overriding objective is similarly hard to identify; perhaps economic security, very broadly defined, comes as close as any. MITI is concerned about raising productivity, strengthening international competitiveness, continually moving up the ladder of value added (while retaining an infrastructure in certain basic industries), achieving efficiency in the use of finite resources, maintaining good relations with major trade partners, and improving the overall quality of life.

For latecomer nations, the goals may include industrial catch-up, protection against foreign economic and political domination, adjustment to changing comparative advantage, and provision for social overhead investments. The inventory of policy instruments used to achieve these and other objectives is fairly standard, including such measures as tax incentives, research and development (R&D) subsidies, restrictions on foreign imports, and so forth.[22]

Because the goals of industrial policy are usually multiple, they can be at odds with each other and with macroeconomic policies. In seeking to maintain high employment levels, for example, a government might place restrictions on foreign imports; but trade protection can undercut other national objectives like the upgrading of manufacturing efficiency and productivity. Because advanced industrial economies are complex and highly differentiated, with specific sectors sometimes seeming to require individualized treatment, the task of weaving diverse strands into a single, systematic design is exceedingly hard to accomplish. Quite apart from the economic obstacles, the difficulties of systematic integration are compounded by political factors—particularly the uneven distribution of power among interest groups, flaws in the structure of institutions charged with aggregating private interests, and high levels of politicization. Politics, perhaps more than anything else, can bedevil efforts at fashioning a rational and effective industrial policy.

Market Failures

In the United States, the government operates on the assumption that it should give the market as much leeway as possible to function efficiently, free from undue interference. Only when the market malfunctions or produces unacceptable results is it necessary for the government to step in and take action. The presumption that the market is the most efficient mechanism of resource allocation is perhaps the first and most basic commandment in the ideology of market capitalism. Of all the major capitalist countries in the world, America is probably the most strongly committed to the gospel according to Adam Smith. The fervor of this commitment runs deep, even though it is not always reflected in actual patterns of government behavior.

Perhaps the clearest illustration of market failure, which has goaded both the Japanese and the American government into taking action, revolves around what economists call "externalities": that is, economic costs or adverse social consequences caused by the failure to anticipate secondary effects. Environmental pollution, involving the discharge of chemical wastes from industrial plants, is perhaps the most notorious example. Without market incentives to maintain clean water and air, the government has had to step in and force industry to comply with standards that protect the public's interests. If it did not, industry might go on polluting in callous disregard for the social costs.

Another example of market failure, less physically harmful to the public but more insidious to the economic well-being, is the problem known as the "collective good" or "public good" phenomenon.[23] In high-tech industries, this problem manifests itself in the unwillingness of private enterprise to bear the costs of basic, nonapplied research—research, that is, yielding no tangible product for commercialization.[24] Since basic knowledge is an inexhaustible public good, the "rational" short-term strategy for companies is to let others assume the costs of basic research and reap the fruits of that spadework by concentrating their own resources on commercialization. This "free ride" mentality is reinforced by the spiraling costs and uncertainties of basic research. If all companies follow this strategy, aggregate investments in basic research—the foundation for breakthroughs in commercial R&D— will fall far short of desirable levels. Everyone will be worse off.

As a nonprofit entity responsible for the collective good, the government is often the only institution capable of preventing this from happening. It alone has the responsibility and authority to underwrite the costs of basic research or to create the tax provisions and legal conditions necessary for profit-conscious corporations to commit substantial resources of their own toward basic R&D. With respect to high technology, governments practically everywhere have allocated large sums to underwrite basic R&D.[25]

National Differences

Although Japan and the United States share a commitment to the market, the two differ significantly in their perceptions of what constitutes market failure and what steps need to be taken to deal with such failure. Japan draws the boundaries of market failure more broadly, including not only obvious breakdowns, like pollution, but also anticipated breakdowns and unacceptable outcomes with respect to what the government deems to lie in the public good and national interest. Even when the market is functioning normally, the anticipated outcome of pure market competition may not be politically palatable; commercial and economic subordination to foreign countries is an example of an unacceptable outcome.

Tokyo's greater disposition to act on the basis of anticipated outcomes sets it apart from Washington, which tends to deal primarily with failures *after* they have occurred. At the risk of oversimplification, perhaps the United States' concerns can be described as reactive, ad hoc, and focused on market failures without reference to industry-specific goals. By contrast, Japan's approach is anticipatory, preventive, and aimed at positively structuring the market in ways that improve the likelihood that industry-specific goals will be achieved. There is a fundamental divergence in expectations and objectives and hence in policy actions. Whereas Americans are content to let the chips fall where they may, the Japanese prefer to remove as much of the element of uncertainty as possible from the market processes. Their disposition to bend, twist, and shape the market is analogous to their practice of using ropes, wires, and strings to bend and twist the trunks and branches of trees into shapes that fit the Japanese aesthetic composition of a landscape, garden, or bonsai plant.

The distinction between reaction and anticipation may appear to split hairs, but it reveals basic differences of far-reaching significance. The United States' reactive approach suggests a preference for leaving the market alone unless there is tangible evidence of a breakdown.[26] This laissez-faire philosophy is premised on an abiding faith in the efficacy of Adam Smith's invisible hand. Japan's preventive approach, by contrast, suggests a more active posture, based on a disposition to steer the market in desired directions. MITI officials are skeptical that a strictly hands-off posture will yield outcomes that coincide with sectoral priorities, public interests, and national goals. To derive optimal outcomes, the visible hand of the state must work in conjunction with the invisible hand of the market.

In the eyes of Japanese officials, the "pure" market is flawed by several shortcomings: imperfect information; narrow, short-term pursuit of instrumental gain; primacy of individual company interests over collective interests; "free ride" approach to the public good; opportunistic behavior; scant spirit of cooperation; structural change and social dislocations; potential subordination to foreign commercial interests; and inattention to national goals. Although the market imparts substantial impetus to long-term economic efficiency, it offers no guarantee that broader social, political, or economic security interests will be served. Because collective objectives are especially important in Japan, a goal-oriented country, MITI officials rely on industrial policy to compensate for the above-mentioned shortcomings in the marketplace.

Antitrust

A classic example of market failure is private-sector collusion in restraint of competition. Such abuses as price-fixing call for continuous monitoring and vigorous enforcement of antitrust laws to safeguard consumer interests and the democratic processes. Antitrust policy is, accordingly, an important facet of the interface between industrial policy and the market.

Outwardly, Japanese antimonopoly laws bear a close resemblance to those in the United States, owing to the historical origins of Japanese postwar statutes in a variety of reforms undertaken during the interregnum of the U.S. occupation.[27] But there are significant differences in the way the codes are interpreted and applied. Stated somewhat simplistically, the Japanese government takes a more pragmatic approach to antitrust enforcement, one that makes allowances for national goals such as industrial catch-up. It takes into account other collective values and extenuating circumstances in weighing enforcement decisions against the letter and spirit of antitrust laws. Included here are such considerations as economies of scale, enhanced efficiency, optimal use of scarce resources, international competitiveness, heightened productivity, business cycle stabilization, industrial orderliness, price stabilization, and economic security. As an industrial latecomer struggling to catch up with Western front-runners, Japan has tended to emphasize producer priorities, economic security, and national interests over doctrinaire adherence to principles of antitrust.

Antitrust is not viewed as an end in itself; it is instead a means of guarding against abuses that might damage Japan's collective well-being. Since antitrust exists in a complicated relationship with other forces in society, business leaders believe that overzealous prosecution of antitrust should not be allowed to override all other considerations; it should not override collective objectives of great national urgency or value. When Japan agreed to liberalization and faced the threat of foreign domination, accordingly, the Japanese government set about "rationalizing" its industrial structure by encouraging mergers between companies in order to bring about greater efficiency in production and to make Japanese producers more internationally competitive. Serious misgivings about the government-led trend toward greater market concentration were expressed during the late 1960's,[28] but the Fair Trade Commission's (FTC) approval of the merger of the Yawata and Fuji steel companies into Nippon Steel reflected the pragmatic accommodation of antitrust policies to the changing circumstances of liberalization.

The government's active role in administering industrial policy complicates the task of enforcing antitrust in Japan. Unlike the Antitrust Division of the U.S. Department of Justice and the Federal Trade Commission, which are concerned solely with reining in or deterring abuses in the private sector, Japan's FTC must also monitor the anticompetitive consequences of MITI's close relationship with industry, its reliance on industrial associations, and the use of administrative guidance concerning investments, exports, and sometimes production. Such forms of government-business coordination are common in Japan. In some cases, administrative guidance may violate antitrust provisions by dampening competition or distorting the price mechanism. This implies that Japanese industrial policy contains within itself the seeds of conflict with antitrust laws. The two exist in a high state of tension.

The Japanese government has found ways of reconciling industrial policy with antitrust enforcement. The FTC and MITI have been able to work out differences of opinion through negotiations. In Japan's latecomer system, the commitment to principles of antitrust has not been allowed to shape the content of industrial policy. There are not very many examples that can be cited in which the FTC has ordered MITI to make major changes in industrial policy in order to conform to antitrust statutes. MITI officials try to take antitrust factors into account in formulating industrial policy.

On the other hand, MITI does not have carte blanche to implement whatever policies it pleases. Antimonopoly enforcement in Japan is not a meaningless charade. Though much less powerful than MITI, the FTC takes its mandate seriously, and the combination of FTC opposition, the aroused opinions of scholars and the mass media, and political pressures from opposition parties establishes definite limits on how far industrial policy can be pushed.

To put Japanese antitrust in comparative perspective, the point should be made that in terms of the stringency of prosecution, the United States is the exception rather than the rule.[29] Japanese attitudes may strike Americans as lax; but compared to antitrust enforcement in France or Italy, Japan's enforcement is far from negligent. Hence, comparisons with the United States may show Japanese antitrust policies in too harsh a light.

Regulation

The distinction between prevention and reaction with respect to market failures can be seen in the different approaches taken by Washington and Tokyo to regulatory policy. In both countries, the government assumes responsibility for shielding the public from dangers to its safety and health; but the two differ in the ways they regulate their economies. In the United States, the objective is to bring the level of risk down to a range deemed acceptable in terms of calculated trade-offs between costs and benefits. Since risks can never be totally eliminated—except at unrealistic cost—the idea is to find a low enough threshold of risk so that damage becomes highly improbable while the costs remain acceptable. This is the optimal cost-to-risk ratio.

In Tokyo, the regulatory philosophy is to push as close as possible to eliminating all risks. Just as the goal of quality control in Japanese manufacturing is to achieve zero defects, the aim of government regulation is to come as close as possible to the elimination of all risks. The approach is more defensive yet aggressive in nature. Government officials in such ministries as Health and Welfare and MITI, bearing ultimate responsibility for regulatory disasters, are not satisfied simply to find optimal cost-to-risk ratios. The regulatory standards they set tend to be tougher and more detailed; for private industry, the time spent and resources invested in meeting those standards also tend to be higher.

If problems occur in the United States, there are mechanisms of correction in the private sector. The individual consumer can file lawsuits to enter claims for liability damage. If the government sets the cost-to-risk ratios too low, the "market" automatically corrects the problem through the ubiquitous instrument of private-sector litigation. Over the years, the number of lawsuits has skyrocketed, and with it the costs of liability insurance, litigation, court awards, and out-of-court settlements. From 1975 to 1985, the United States witnessed a 1,000 percent increase in product liability cases brought before the federal courts and a 401 percent increase in liability awards.[30]

Needless to say, this has had a decidedly negative impact on the willingness of U.S. companies to assume risks in developing and selling certain products with high liability thresholds, like oral contraceptives and certain types of vaccines. The metastasis of litigation (not to mention the mere threat of a lawsuit) has had the effect of tightening the regulatory safety nets—but at extraordinarily high cumulative costs in terms of commercial risk-taking and entrepreneurship.

Japan has not had anything like this frenzy of lawsuits. It is not a litigious society, and the courts are not the main mechanisms of conflict resolution. This means that government officials may be under heavier pressure to safeguard the public than their counterparts in the United States, because there is less leeway for citizens to seek compensation through legal channels. Ultimately, it is the responsibility of higher civil servants to set regulatory standards that are sufficiently foolproof to protect the public from daily damage and occasional disasters.

To do this, they must consult closely with private-sector producers in ways that ensure voluntary compliance, if not positive cooperation. Regulatory control is not achieved through unilateral decree, backed by the threat of legal sanctions, as it is in the United States. It is achieved, like most other public policies, through consultation, consensus, and voluntary compliance. What makes the Japanese approach so interesting is that it makes government intervention seem less coercive than in the United States, yet it still manages to set regulatory standards that are as high.

In certain areas of regulatory control, the visible hand of government is much heavier in the United States. The federal government wields a much stronger hand in areas of social justice, such as affirmative action and equal opportunity employment, and in activities related to national security, such as export controls over dual-purpose technology (technology for both civilian and military use), military specifications, and classification of information related to national security. Owing to its social homogeneity and low military posture, the Japanese government has not had to play a very intrusive regulatory role in these areas.

The military responsibilities alone of the U.S. federal government are so burdensome that it has not been able to adhere to the kind of hands-off policy that faith in the invisible hand of the market would dictate. The swath cut in the market economy by intervention driven by national security has had a major impact on the development of sectors like high technology. It also has altered the structure of the market economy as a whole and strengthened the state's overall role.

Market Organization

Variations in state intervention can be attributed, in part, to differences in the way the markets in various countries are organized, as well as to differences in ideology.[31] Indeed, a country's market structure is a central determinant of industrial policy, because it provides the framework within which concrete measures are formulated and applied. Peter Hall argues that Britain's economic decline can be attributed primarily to defects in the structure of its market, which public policy failed to correct.[32] The intensity of market competition, the kinds of market failures that occur, how serious they are, and what industrial policy responses are required all emerge from the crucible of organized markets.

In Japan, capital and labor markets are organized along very different lines from those in the United States and Great Britain. Based on the norm of career-long employment, the Japanese labor market is less mobile and more seniority-oriented, with greater emphasis placed on in-company training. As discussed in Chapter 3, the equities market is less fully developed than those in the United States and Britain. In consequence, Japanese government officials have seen both the need and the opportunity to intervene more often, across a broader expanse, and in a greater variety of ways. Unlike its U.S. counterpart, the Japanese government has never felt comfortable about letting the decentralized market take its "natural" course. The structure of its capital market has not been developed enough to allocate capital with optimal efficiency or in keeping with sector-specific goals.

Seen in the context of its market structure, therefore, Japanese industrial policy has fulfilled some of the functions that fully developed and self-regulating markets have served in other countries. Indeed, Gary Saxonhouse has argued that Japanese industrial policy can be understood only in terms of its role as a compensating mechanism for market shortcomings.[33] To make up for structural deficiencies in Japan's equities markets, for example, the process of identifying industrial priorities has sent market-type signals to financial institutions that, in the past, have acted on such signals to allocate capital to priority sectors. Likewise, owing to bottlenecks in labor markets, the publication of government white papers on future industrial structure has communicated clear signals to college graduates, prompting the best and the brightest to seek employment with blue-chip companies in the fastest-growing sectors. Japanese industrial policy can thus be seen, in some senses, as a compensating mechanism for imperfections in capital and labor markets.

The instruments of industrial policy are also designed to treat problems that emerge from special market conditions in Japan. The vulnerability of highly leveraged companies to fluctuations in business cycles and the dependence of subsidiary and subcontracting networks on parent corporations, for example, have prompted MITI to administer industrial policies designed to decrease some of the risks of cyclical downturns. Failure to lower the risks would lead to unacceptably high levels of bankruptcies and unemployment. Thus, cyclical vulnerability is the market context within which antirecession and rationalization cartels and special financing for small and medium-sized businesses have been administered.

At the same time, as Chapter 3 points out, Japan's market structure provides multiple points of entry for government intervention. In particular, the existence of what might be called "non-market" or, perhaps more accurately, "extra-market" institutions—*keiretsu* structures, extensive patterns of intercorporate stockholding, close banking-business ties, subsidiary and subcontracting networks, specialized trading companies, and industrial associations—presents MITI with multiple points of entry through which to exert a direct influence on market outcomes. MITI can affect developments in the market by quietly leaning on large companies, industrial associations, banks, securities houses, and other private-sector institutions.

The prevalence of extra-market institutions permits MITI to bend and shape the market without having to go through the political channels of the LDP and the parliament. From the standpoint of implementing industrial policy, the existence of direct channels of influence is an asset of enormous (if overlooked) importance. It gives MITI ways of circumventing political and legislative entanglements. It also makes consensus formation and administrative guidance effective tools for state intervention.

Such access points are not as prevalent in the United States; but since the U.S. government is not as likely to intervene, the paucity of entry points is not a handicap. In other countries that rely heavily on industrial policy, like France, the organization of economic activity, particularly government control over the financial system, provides similar mechanisms of leverage.[34] Of course, the leverage made possible by the combination of organizational access points and a smoothly functioning market greatly facilitates the state's capacity to administer industrial policy. Where such organizational structures are absent, government-industry interactions are apt to be more distant, antagonistic, and legalistic. Industrial policy is harder to implement.

The intrusion of organizational elements, to be sure, runs the risk of stifling market competition and lowering overall levels of efficiency and this, in turn, may require state intervention to correct.[35] Extra-market mechanisms, like long-term relationships based on obligated reciprocity, can give rise to rigidities that threaten to disrupt the smooth functioning of the market. In certain areas of the Japanese economy, like retail distribution, structural rigidities have created enclaves of inefficiency that have imposed costs on consumers and given rise to trade friction with foreign producers. The potential downside costs, stemming from the pervasiveness of organizational structures, are thus considerable. Nevertheless, as Ronald Dore points out, most of the serious pitfalls have somehow been avoided.[36]

The very existence of extra-market structures also increases the temptation to intervene. During the early years of catch-up, MITI made extensive use of the private-sector channels of access to exercise what were then its formidable instruments of power. Interestingly, however, MITI has seldom used all levers of power at its disposal, even during the heyday of high-speed growth. Its pattern has been to intervene forcefully but selectively on the basis of prior consultation with the private sector, and then to pull back. Seldom has the intervention led to a permanent ratcheting upward of state power.

Areas in which MITI has actively intervened include: (1) consensus building and the articulation of a long-term "vision" for those industries under its jurisdiction; (2) the setting of sectoral priorities; (3) the allocation of subsidies and facilitation of financial flows to priority sectors; (4) adjustments of industrial structure; (5) infant industry protection; (6) investment guidance in certain industries and under certain conditions; (7) regulation of excessive competition; (8) downside risk reduction and cost diffusion; and (9) export promotion and mediation of trade conflicts. Very few of these activities fall within the sphere of what the U.S. government considers its proper responsibility. In consequence, the U.S. government has refrained from intruding into these areas of the market economy. Nowhere is the contrast between Japanese and U.S. approaches brought more graphically to light.

Macroeconomic Policy

The United States' restrained approach to industrial policy is based not only on its ideological commitment to market forces but also on its preference for relying more heavily on macroeconomic measures. The best remedy for problems in individual sectors, many U.S. leaders feel, is to invigorate the entire economy through sound monetary and fiscal policies.[37] American economic and political institutions are not particularly well suited to produce a differentiated set of policies for specific sectors. In the United States, any attempt to fine-tune sectors is apt to lead to greater levels of politicization, and this has the effect of doing more harm than good.

Japanese economic and political institutions make it easier to engage in sectoral fine-tuning; the spheres of industrial and macroeconomic policies are less disconnected. Although the two are not deliberately integrated, some effort at coordination is made. Tax measures and MITI expenditures, for example, are both subject to strict guidelines established by the Ministry of Finance (MOF). In compiling a budget each year, MITI and MOF exchange a great deal of information and engage in continuing negotiations over line items. Such negotiations can be considered a form of conscious coordination. Many industrial policy measures could not be undertaken by MITI without MOF's explicit approval. MOF is very much a partner in the industrial policy formulation process.

In formulating industrial policy, MITI must also take macroeconomic conditions into account, because investment decisions and corporate strategies are sensitive to fluctuations in interest rates, money supply, and exchange rates. Because authority over monetary policy is vested in the hands of the MOF and the Bank of Japan (BOJ), Japanese industrial policy must be conducted within the broad, macroeconomic parameters set forth by these two financial agencies. Although it would be an exaggeration to call the communications and negotiations that take place between MITI and MOF a form of "integrated" or "coordinated" planning, industrial policy is not formulated in total isolation from macroeconomic policy. At the very least, MITI must see that industrial policy does not work at cross-purposes with MOF's macroeconomic orientation. Imperfect though the coordination is, it goes beyond what most other industrial states have managed to achieve.

It is possible to argue, as most economists do, that Japan's postwar success owes more to the soundness of macroeconomic policies than to its industrial policy.[38] This is certainly a plausible argument, one that can be convincingly stated. But without entering into the pros and cons of the argument, one can propose a variant hypothesis: sound macroeconomic management greatly facilitates—and may in some senses be a condition of—effective industrial policy.

Conversely, ineffectual macroeconomic management exacerbates the difficulty of the tasks falling to industrial policy. Serious problems at the aggregate level usually translate into pressures for special relief in troubled sectors. But at precisely the time when interest-group demands peak, the capacity to deliver technically sound industrial policies (representing something more than short-term, political concessions) may be at its lowest. Double-digit unemployment, for example, can have the effect of constricting the range of viable policy options. Overburdened by interest-group demands, the political system itself may malfunction. Once a series of bad industrial policy decisions have been made, the damage is hard to reverse.

The weight that has fallen on industrial policy is much heavier today than it was before the first oil crisis (1973–74). This is due to the confluence of several global developments: (1) the dramatic proliferation of international linkages, particularly in the area of foreign trade; (2) the scope and speed of domestic adjustments necessitated by such interdependence; (3) stepped-up competition among industrial states; (4) growing signs of trade protection and strains in the GATT-based international trade regime; (5) problems connected with the shift to floating and flexible exchange rates; (6) global recession, caus-

ing countries to struggle with problems of sluggish growth, high inflation, and high unemployment; and (7) the seemingly insoluble
problems of working out multilateral coordination of macroeconomic
policies.[39] Needless to say, these and other developments have placed
all industrial states in a serious economic bind.

In this sense, the effectiveness of macroeconomic policies in Japan
has lightened the burdens that might otherwise have fallen on industrial policy.[40] Furthermore, unlike the United States and some European countries, Japan has not yet had to face the problem of coping
with the obsolescence of entire smokestack sectors. Except for shipbuilding, aluminum, and a few other industries, old-line manufacturing sectors have managed, by and large, to stay competitive in world
markets. According to standard indicators of economic performance,
such as productivity gains, inflation and unemployment rates, and
capital investment, Japan has done at least as well as other advanced
industrial states.[41] In consequence, the government has not had to use
industrial policy as much to rescue declining industries, bail out firms
on the verge of bankruptcy, or force "voluntary" export restraints on
trade partners.

For structurally depressed industries like shipbuilding and aluminum, the Japanese government has tried to devise policies that reduce
overall capacity in order to phase out obsolescent plants and inefficient producers.[42] Although aspects of Japan's adjustment policies—
like the formation of rationalization cartels—have engendered foreign
objections, the Japanese insist that their policies are designed to avert
protectionism by adapting industrial structure to shifts in the international division of labor. Japanese industrial policy for structurally
troubled industries, as MITI officials portray it, follows the natural
flow of long-term market forces—unlike policies in the United States
and Europe, which attempt to rescue declining sectors from the rising
currents of international competition.

Latecomer Catch-up

Industrial Catch-up

One of the most decisive forces shaping Japanese industrial policy
has been the country's historical experience as an industrial latecomer.
This has had a far-reaching impact on virtually all aspects of the
Japanese rationale for market intervention. From the mid-nineteenth
century until the mid-1970's, Japan's overriding goal was to industrialize as fast as possible in order to catch up with and overtake the

leading powers of the West.[43] For three-quarters of a century (1868–1945) following the Meiji Restoration, the Japanese government was consumed by the task of constructing the industrial infrastructure necessary to build and sustain a world-class military machine. What haunted Japan from the start was the specter of Western colonization and foreign domination.

Following its crushing defeat in the Second World War, Japan renounced the military goals that had propelled it to become the first non-Western nation to industrialize. In place of an instrumental strategy—industrial strength for conversion into military power—postwar Japan sought economic prosperity for its own sake. And having decided to re-embark on the road to industrialization, postwar Japan had to resume the catch-up race, a task of utmost urgency because of the extensive damage done by incendiary bombing to the country's industrial infrastructure (estimated at two-fifths of the nation's capital stock).

The decision to try to join the select circle of advanced industrial powers may have seemed overambitious, given the resource limitations of Japan's factor endowments. In particular, the strategy of concentrating on capital- and technology-intensive industries may have seemed illogical—a needlessly difficult way to begin—considering the comparative costs of production and Japan's low wage advantages. A labor-intensive strategy, based on light manufacturing, might have been the path of least resistance.[44]

But the more ambitious and challenging route through heavy manufacturing had much to recommend it. From a long-term perspective, capital- and technology-intensive industries offered an assortment of advantages—high income elasticity of demand, big spillover effects, the potential for very rapid technological advancement, opportunities for steep gains in labor productivity, an infrastructure capable of sustaining Japan's large labor force at full employment, and the prospect of someday reaching a standard of living comparable to that of the advanced countries in the West. In short, the route through heavy manufacturing held out the promise of very rapid economic growth within the framework of an enduring economic infrastructure; the creation of such an industrial base could support the Japanese for generations to come, even if it involved higher initial costs and risks.

To industrialize in as short a time as possible called for extensive government involvement. Specifically, the government felt a responsibility to step in to perform such key functions as delineating a long-range set of goals, drawing up a series of medium-range plans, mobilizing scarce resources, responding to the particular needs of individual sectors, and dealing effectively with overseas markets and resources, not to mention facilitating foreign trade. It required, in short, the implementation of industrial policy, as Chalmers Johnson points out.[45]

Although Japan faced no external threat as it had before the war, government leaders felt compelled to preside over the reconstruction of its war-razed economy. Suffering from an acute shortage of capital, with inadequately developed mechanisms for capital allocation, the Japanese economy had to be made to produce enough value added to support a large population on a small and crowded archipelago lacking in nearly all natural resources. The urgency of the task and the devastated condition of the private sector prompted the government to take charge from the very beginning.

The situation, though discouraging, was not entirely bleak. The Japanese are adept at converting national crises into a driving sense of national mission. In the aftermath of defeat, the pressing need to cope with dire economic circumstances supplied that kinetic sense of national purpose. Almost as a by-product, it also provided the rationale for vigorous government intervention through the vehicle of industrial policy.

During the 1950's and 1960's, industrial policy played a central role in meeting the historic challenge of industrial catch-up (a role that has shrunk with the passage of time, as Japan has closed the gap). Showing little faith in the magic of Adam Smith's invisible hand, the Japanese government intervened in order to (1) establish sectoral priorities; (2) mobilize resources to hasten their development; (3) protect infant industries; (4) issue guidance on investment levels; (5) organize rationalization and antirecession cartels; (6) allocate foreign exchange credits; (7) regulate technological flows in and out of Japan; (8) control foreign direct investment; (9) issue "administrative guidance" (*gyōsei shidō*) enjoying quasi-legal status; and (10) publish white papers on mid- and long-term visions of Japan's future industrial structure.

Faced with the exigencies of industrial catch-up, the Japanese government clearly felt that it could ill afford the luxury of letting its economy evolve slowly and "naturally." Market forces could not be allowed to take their own course; they had to be harnessed, directed, and spurred on. In its perception that the state had no choice but to play a vigorous role (if certain national goals were to be reached), Japan displayed some of the classic characteristics of latecomer states, straining mightily to step out of the shadows cast by a few industrial behemoths that dominated the world economy.

Industrial Planning

Latecomer catch-up called for some degree of government guidance, but not the imposition of central planning commonly associated with socialist or dirigiste states. Notwithstanding images of "Japan, Inc.," the Japanese economy has never been, and is in no meaningful sense today, centrally planned.[46] Although the government issues medium-term (3–5 years) and long-range (5–10 years) plans on a regular basis, these plans are in no way binding either on the private sector or on decisions made within the government itself.[47] They serve only as general forecasts, not authoritative directives. The actual performance of the economy invariably follows a divergent trajectory.[48] Even short-term plans of one year or less do not always turn out to be accurate predictors of economic performance, though the shorter the time span, the closer the convergence between forecast and outcome.

Although economic planning plays a smaller role in Japan than it does in France, it assumes a bigger role than in the United States, where the very idea of government planning is considered almost sacrilegious. Because the Japanese are compulsive about minimizing uncertainty, the government tries to draw a general road map for the market economy to follow. Specifically, it seeks to identify concrete, near-term goals within a long-term "vision" of desirable future directions for the industrial economy. Since such plans emerge from close, ongoing consultations with the private sector, they do not have to be forcefully imposed. Their main value is the promulgation of an orderly frame of reference within which corporate actors can make a series of their own decisions. When economic plans are concrete, industry-specific, and based on a consensus between MITI and industry, they can get actors in the private and public sectors moving in the same direction.

From the end of the war to the mid-1960's, Japanese officials were worried that, without an explicit set of allocative priorities, scarce resources might be diverted from electrical power generation and steel to sectors of secondary importance, like coffeehouses and real estate. Hence, MITI and industry have subscribed jointly to the notion that it is necessary for the government to set broad directions and establish clear priorities.

Table 1.1 outlines some of the major industrial policies followed by the Japanese government during the 1950's and 1960's, an era of unprecedented catch-up and growth. As Table 1.1 indicates, the scope of Japanese actions far exceeded that of the U.S. government's in terms of initiatives taken to structure the market. The emphasis was placed on building a sturdy infrastructure of heavy industry and on making the most efficient use of scarce resources.[49]

From the mid-1960's to the mid-1970's, while continuing to pursue the goal of industrial catch-up, the Japanese government turned its attention to the task of dealing with the potentially far-reaching effects of trade liberalization.[50] With Japan fully recovered from the war and growing faster than any other industrial country, foreign governments demanded that the walls of infant industry protection be brought down. In stages, the Japanese complied, but at a sluggish pace and often in ways that merely shifted the barriers from formal (tariffs and quotas) to informal obstructions (customs and inspection red tape, national standards, and buy-Japanese preferences). Here was a case where the application of foreign pressure succeeded in forcing Japan to make changes, but the changes could be neutralized by informal barriers in the private sector. Policies changed but outcomes did not.

To compete against giant foreign producers, MITI took the lead in pushing for structural "rationalization" in a number of key industries, like steel and automobiles. This push for rationalization took the form of encouraging mergers among various domestic competitors so as to increase returns to scale and prepare Japanese firms to take on the foreign Goliaths. The merger of two large manufacturers, Fuji and Yawata, into the world's biggest steel producer, Nippon Steel, was the most spectacular example of Japan's industrial policy response to the pressures of trade and investment liberalization.[51] Except for Nippon Steel, however, MITI's efforts at structural rationalization met with limited success. Not only did MITI fail to anticipate deep-seated private-sector resistance, it also underestimated industry's capacity to compete, even in a liberalized environment.

Long Production Runs and Process Technology

For latecomers like Japan, which pursue the heavy manufacturing route to industrial catch-up, perhaps the two keys to success lie in achieving large economies of scale and improving process technology. Mass production permits the latecomer to lower the per-unit cost of production, invest in new plant capacity, implement a strategy of import substitution, compete vigorously in export markets, and develop new and better process technology. Japan has had the good fortune, which may latecomers lack, of having a large and rapidly

TABLE 1.1

Postwar Japanese Policies for Industrial Catch-up, 1955–1973

Overall directions and goals
  *Long-term visions for Japan's industrial economy
  *Special development laws for priority industries
  *Annual delineation of goals for each industry, based on consensus between MITI
    and industry

Industry promotion measures
  Tax incentives
    Accelerated depreciation allowances
    Research and development
    *Nontaxable reserve funds for retirement compensation and special contingencies
  Financing
    *Japan Development Bank loans
    Other government financial institutions
    Long-term credit banks
  Industrial structure
    Preferences for long production runs and market concentration
    *Rationalization: encouragement of mergers
    Investment and production guidelines
    Assistance for small and medium-sized enterprise
  Risk reduction
    *Antirecession cartels
    Public procurements
  Technological development
    *Identification of national priorities for technological development
    *Encouragement and assistance in advancing manufacturing and process
      technology
    Control over technology licensing
    R&D subsidies
    Government-sponsored research projects
    Basic research conducted at government laboratories
    *Administrative guidance
    Flexibility for ad hoc responses to developing situations
  *Antitrust
    Selective exemptions
    Flexible enforcement
  Manpower policies
    *Demonstration effect: patterns of recruitment in keeping with industrial policy
      priorities
  Industrial location
  *Coordination of industrywide consensus
    *Close working relationship between MITI and industry
    *Industrywide aggregation through industrial associations

International interface
  *Infant industry protection
    Tariffs and quotas
    Nontariff barriers
    *Control over foreign direct investment (until 1970's)
  *Allocation of foreign currency exchange to specific sectors, industries, and
    companies

TABLE 1.1 (cont.)

International interface (continued)
  Export promotion
    Special tax deductions for exports
    Overseas market information
    Export-import financing
    Mediation of trade friction
  Economic security
    *Raw materials procurement and stockpiling
    Foreign aid
  Multilateral activities
    Participation in international organizations
    International capital flows
      Exchange rates
      Overseas investments

---

*Denotes differences in kind or degree from U.S. government policies.

growing domestic market. The luxury of having a large domestic market can scarcely be overstated, especially if it can be closed to foreign entry through infant industry protection.

On top of this, Japan had relatively free access to overseas markets, which multiplied the returns to scale. This combination—captive domestic demand and accessible overseas markets—gave Japanese old-line industries significant cost advantages in world competition, even after they lost their early edge in low wages. Latecomers failing to achieve volume production have a hard time keeping a competitive edge, once wages begin climbing to levels nearer those of fully industrialized states.

Japanese companies were also able to upgrade process technology, thanks to massive imports of foreign know-how,[52] adaptations and incremental improvements, effective subcontracting networks, and support provided by Japanese industrial policy. MITI placed enormous emphasis on upgrading the country's overall capabilities in process know-how. Not that companies had to be prodded; because an advantage in process technology could be converted into significant gains in market share, Japanese companies were eager to move down the learning curve of long-run production as swiftly as possible. The enticement of expanding market shares supplied more than enough incentive.

The policy of prohibiting foreign direct investment in Japan, begun under the U.S. occupation and continued through much of the 1960's, had the effect of encouraging foreign companies to license patents, since direct manufacturing in Japan had to be ruled out. Wanting to facilitate improvements, MITI monitored domestic and overseas developments in process technology and, from time to time, stepped in to help Japanese manufacturers negotiate with foreign companies for licensing rights. Other features of the Japanese industrial system—subcontracting networks, permanent employment, and corporate emphasis on production engineering—also contributed to national advances. Thus, large Japanese corporations functioned within an industrial system admirably well suited to meet the needs of latecomer development.[53]

The speed of Japan's industrial growth, the prime objective of its industrial policy, was astounding. By 1970, Japan had actually overtaken foreign front-runners in many manufacturing sectors, such as shipbuilding and steel. In spite of its spectacular successes, however, Japan has not completely shaken its underdog mentality; nor has it abandoned many of the industrial policy tools that were used during the heyday of postwar catch-up.

Where Japan has changed is in the composition of its industrial structure and its redefinition of national needs and goals. Since the mid-1970's, the focus of attention has shifted from liberalization to structural transformation—from energy-intensive heavy manufacturing to knowledge-intensive information industries. It is, in essence, the transition from latecomer to pioneer.

End of Latecomer Catch-up: Persistence of Crisis Psychology

In spite of the fact that Japan has closed the gap in all but a few sectors, the sense of national urgency is as strong today as it was during the 1950's and 1960's. Why? Because so much is at stake, and the pace of commercial and technological change is so fast that stragglers cannot afford to fall further behind. In most high-technology industries, learning curves tend to be steeper than those in old-line manufacturing, and product life cycles much shorter. This means that the costs of lagging behind can be significantly higher. The sense of urgency is compounded by the size and visible power of foreign giants in certain high-technology fields—like Boeing in aircraft, AT & T in telecommunications, and IBM in computers. The Japanese feel they must muster as much effort in high technology as they did in old-line manufacturing or risk domination by foreign Goliaths.

There may be no compelling economic reason for Japan's preoccupation with the threat of foreign domination. According to the Ricar-

dian theory of comparative advantage, the strength of foreign producers in one field means that different fields of comparative advantage
will be left open.[54] Unless there is cause to doubt the reliability of
foreign suppliers, or disgruntlement about the specialized niches available through the international division of labor, the fact that IBM,
instead of Nippon Electric Company (NEC), produces computers
should not be grounds for excessive concern. However, for reasons
that can only be regarded as curiously nationalistic, MITI and the
electronics industry have been and remain deeply worried about the
implications of IBM's domination. Staving off domination by IBM,
even eventually overtaking IBM if possible, may be the overriding
objective for Japan's computer industry.

The reason for this preoccupation resides in the natural instinct for
survival in what some Japanese see as a marathon race for economic
pre-eminence. The information industry, MITI officials believe, holds
the key to future competitiveness across a whole range of industrial
sectors—not only high technology (robotics, machine tools, telecommunications) but also the old-line industries (steel, automobiles, chemicals) and even the services (banking, insurance, distribution). Much
more is riding on the competition than leadership in a single sector. If
Japan fails to meet the foreign challenge, many Japanese fear their
country will be forced to the periphery of economic activity, doomed
to subordinate status as an industrial has-been.

To non-Japanese, such scenarios seem overdrawn. Is it realistic to
assume the worst possible outcome—economic decline and international subordination—if Japan falters in one or two areas of high
technology? Nonetheless, such fears—real, imagined, or deliberately
cultivated—motivate the Japanese to step up their efforts to catch up
with, and if possible overtake, leading foreign giants like IBM. Here,
in short, is another example of Japan's catch-up mentality and the use
of a sense of external crisis in mobilizing for collective action.

The International Dimension

Economic Security

Japan's race to catch up with the West suggests that it has had a
deep-seated preoccupation with its economic security, a subject often
overlooked in foreign discussions of Japanese industrial policy but of
fundamental importance.[55] Japan's concern for economic security is
certainly understandable, given the country's almost total lack of raw
materials, its geographic isolation, and the trauma of its prewar ex-

periences. The desire for economic security is manifest in four areas of industrial policy: the procurement of raw materials; levels of international interdependence; access to export markets and foreign technology; and overall industrial structure.

The Japanese government has always wrung its hands over the availability of raw material supplies. The shock of the first oil crisis (1973–74) did nothing to alleviate its fears. Acting to safeguard Japan's economic lifeline, MITI has put together or coordinated a large number of overseas projects involving resource extraction and upstream development as well as private-sector agreements for long-term purchase.[56] International projects or resource-related agreements that contribute to Japan's economic security often qualify for low-interest government financing through the Japan Export-Import Bank. If the project is designated a "national project," it can also qualify for a guarantee against losses.[57] Also falling within MITI's jurisdiction is the coordination of efforts at stockpiling such essential raw materials as oil, as a hedge against unforeseen crises.

With MITI's assistance, Japanese companies have concluded a large number of agreements with resource-rich countries, particularly in the wake of the first oil crisis, when anxieties ran high about the availability of long-term supplies of essential raw materials. Indeed, Japanese trading companies, banking consortia, and private corporations (especially those involved in resource extraction) have gained a worldwide reputation for effective coordination, painstaking fact-finding, and determined pursuit of overseas opportunities, as well as very aggressive financing.

If there is an area of industrial policy where the concept of "Japan, Inc." comes close to reality, it is the area of resource security. Included in this category are all aspects of acquiring essential raw materials: upstream extraction, purchase, stockpiling, and resource-related trade and investment. Of special importance is the security of energy supplies and their conversion into power generation. The Japanese government wields a heavy hand in all facets of essential raw materials acquisition and upstream processing.

A related aspect of economic security is Japan's reliance on foreign sources of technology. One reason the Japanese are so eager to move beyond the frontiers of knowledge and develop their own capabilities in state-of-the-art technology is that they fear that overseas technology may not be readily available in the future. Even under the most favorable circumstances, Japan will have to offer advanced technology in order to obtain technology in return. There may be a groundswell of technological nationalism abroad, arising from the escalating costs and risks of research and development and the severity of commercial competition.

The fear of technological nationalism even extends to the United States, Japan's closest ally, and traditionally the most open source of advanced knowledge in the world. For reasons of national security, the transfer of technology may be increasingly subject to constraints. To lower the level of dependence—already very high in aircraft, space, and other areas of complex systems integration—the Japanese government feels that it must push hard to advance Japan's indigenous capacity to innovate.

Economic security also means continued access to large overseas markets, particularly the U.S. market, which accounts for nearly 40 percent of Japan's total exports. The groundswell of protectionist sentiment in the United States and around the world is deeply worrisome to the Japanese. One of the heaviest burdens to have fallen on MITI's shoulders since the mid-1970's is the task of damage limitation in trade disputes; this involves working to head off conflicts before they explode, or trying to contain the damage if they do.

MITI has concentrated trade-related efforts in three areas: restraining the flood of Japanese exports, opening Japanese markets to foreign imports, and stimulating domestic demand to reduce reliance on overseas markets. Of the three, probably the most effective to date in terms of assuaging foreign resentment has been the acceptance of export restraints, although the expansion of domestic demand may turn out to be of greater long-run significance. At the time of the United States–Japan trade friction over automobiles in the late 1970's, MITI officials went to great lengths to persuade Japanese automakers to accept voluntary export restraints.[58]

An even heavier responsibility has been that of overseeing the transition of Japan's industrial structure from heavy manufacturing to one based on low-energy-consumption, knowledge-intensive industries. Shifting away from heavy manufacturing not only reduces Japan's dependence on overseas raw materials but also is in keeping with trends in international comparative advantage. Economic security is thus a factor of prime importance, constituting grounds for legitimate government intervention in the marketplace. Fortunately for Japan, the government has not had to bail out faltering industries, like steel, deemed essential to its economic security.

Over time, the definition of what constitutes economic security may expand and come to entail starker trade-offs between security and

efficiency, particularly if the external security environment deterio-
rates. National security can be a highly elastic category of government
intervention, as seen in the United States. The Japanese government's
willingness to absorb the costs of economic inefficiency for the sake of
national security is vividly seen in its insistence on producing its com-
bat aircraft (F–15 interceptors and F–16 fighters) domestically under
foreign license—in spite of the fact that this often costs twice as much
as importing them directly from foreign manufacturers (which can
make them much more cheaply because of their longer production
runs).[59] Whether military considerations intrude increasingly into Japan's security picture—with high trade-offs in terms of economic inefficiencies—bears watching.[60]

International Interdependence and Vulnerability

Closely related to economic security—indeed, an integral aspect of
it—is Japan's growing integration in the international system and the
Japanese perception of vulnerability. What happens in international
money and foreign exchange markets, for example, has a substantial
bearing on Japan's domestic economy—just as Japanese financial markets and capital flows influence the volume of transactions that take
place in New York and London. Although Japan is the second-largest
economy in the world, and possesses far greater leverage than it did in
the past, the perception of vulnerability persists.

The feeling was especially strong during the early postwar period,
when Japan was small and fragile; it remains surprisingly strong today
because of Japan's history of isolation from the rest of the world, the
bitter experience of the Second World War, recognition of the fact that
Japan has no natural allies, the country's distinctive culture and society, and, curiously enough, its postwar economic success.

As in the case of the small European states, Japan's perception of
vulnerability has reinforced the sociocultural preference for consensus
as the operative norm of policy-making and reliance on continuous,
informal bargaining between political and economic groups as the
standard mode of conflict resolution.[61] Consensus, discussed more
fully in Chapter 3, carries the weight of a powerful, binding norm in
Japan's policy-making process. Its impact is far-reaching, going beyond the mere policy-making process and extending into the substance
of industrial policy. The consensual mode of interaction between
MITI and private business is a central reason for the pro-producer
(rather than pro-consumer) orientation of Japanese industrial policy.

The sense of international vulnerability has also strengthened the resolve of Japanese higher civil servants to impose structure, order, and sector-specific goals on what would otherwise be freewheeling market forces; this, they feel, reduces the risks and costs of international exposure. At the same time, however, the rising level of international interdependence has persuaded MITI bureaucrats that limits must be placed on the scope of market intervention. Japan cannot afford to intervene in ways that defy market forces; the economic and political costs would be unacceptably high.

The international factor has thus given rise to, and actually helped to shape the content of, industrial policy. In this respect, Japan is more like the small European states than the large industrial powers. In pursuing an active, consensual, yet market-conforming industrial policy, as Peter Katzenstein points out, Japan bears some striking resemblances to Switzerland.[62] Small size and the perception of vulnerability certainly described Japan during the 1950's, but it is interesting that as Japan has become the world's second-largest economy, the emphasis on an adaptive, accommodating industrial policy has persisted. Whether Japan seeks to exert greater influence in the world economy or, alternatively, reduces the role of industrial policy will bear watching in the future.

Adapting Market Forces

The orthodox view in the United States is that the decentralized market, not the central government, is the best and most impartial judge of industrial "winners" and "losers." In Japan, too, the many virtues of market systems are recognized. In addition to its impartiality and efficiency, the market is unsurpassed in its capacity to apply instrumental rationality to virtually all spheres of economic life, from the individual consumer and private household to the profit-conscious corporation. What the market offers is a structure of incentives more compelling than any other ever devised, incentives based on the powerful appeal of rationally calculated self-interest.

Moreover, the network of individual decision-centers through which the market functions generates an extensive circulation of information on the basis of which rational decisions can be made. The heavy flow of information makes market economies less prone to error than centrally planned systems that rely on centralized decisions made by a small cabal of elites. Market systems also possess superior feedback mechanisms, which permit self-correction to take place when mistakes are made.

The superiority of market economies rests, in short, on the myriad of decentralized networks that pump massive amounts of information through all parts of the system. Since there is an inverse relationship between centralization and coercion, on the one hand, and the density and quality of available information, on the other, it is not surprising that market systems function more efficiently than centrally planned economies. More and better information is available, and the economy is relentlessly swept along by the rational pursuit of self-interest; in centrally planned systems, on the other hand, economic behavior is based on limited amounts of information and on directives imposed from above.

One of the most telling criticisms directed against the adoption of industrial policy is that it constricts and distorts the circulation of vital information on which a vigorous market economy is based. When policy-making is excessively centralized, as it often is when industrial policy is adopted, the upward flow of information from lower and middle levels tends to be choked off.

Because Japanese society is organized in ways that diffuse power and decentralize decision making, the information costs of adopting an industrial policy are minimized. Indeed, as pointed out in Chapter 3, the distinctiveness of the Japanese system of industrial policy-making is that it does not constrict the vital flow of information. Leaders in the public and private sectors make a point of exchanging large amounts of information.

By drawing information from a wide cross-section of society—banks, academia, the mass media, industry, and labor—Japan's system of industrial policy-making averts the danger of making major errors in judgment owing to bottlenecks in information flow. "Viewed as a system of information collection and dissemination," Ryūtarō Komiya writes, "Japan's system of industrial policies may have been among the most important factors in Japan's high rate of industrial growth, apart from the direct or indirect economic effects of individual policy measures."[63] Indeed, some Japanese believe their system may be superior to that of even the decentralized market, if one judges in terms of the quality and quantity of information circulated and digested.[64]

For all its virtues, however, Japanese officials are ambivalent about aspects of market capitalism that do not fit neatly into the fabric of Japanese society and culture. Take, for example, the concept of consumer sovereignty, which lies at the heart of neoclassical economics. According to this concept, individual welfare is best served when the individual is free to make his own choices about how to spend his resources to maximize what he values (that is, his utility preferences).

In some respects, the concept of consumer sovereignty seems to have gained acceptance in Japan. "The consumer is king" is an aphorism that Japanese companies have taken to heart. In spite of concerns that consumers may act on the basis of inadequate information and shallow values, the government has not stepped in to provide fuller information on matters vital to the public's health and safety. Unlike the U.S. Surgeon General's Office, for example, the Japanese government has not issued warnings to the public about the health hazards of cigarette tar and nicotine. Nor has it tried to change the utility preferences of individual consumers in ways that might contribute more effectively to social welfare. If there are paternalistic instincts in the Japanese government, they are overwhelmed by a conviction that it is not the proper place of government to tamper with consumer preferences. For better or worse, the Japanese government acknowledges that the consumer is, and should be, the ultimate judge of individual welfare.

On the other hand, the Japanese feel uneasy about the concept of consumer sovereignty, because its implications for Japanese society are not entirely positive. What is most troublesome is the centrality accorded the individual. The individual—his preferences, interests, goals, and legal rights—sits at the heart of the doctrine of consumer sovereignty. The elevation of the individual is foreign to Japanese society, where everyone is subsumed within a collectivity, be it family, firm, or nation. The interests and needs of the collectivity take priority over those of individuals. This is a fundamental principle of social organization.

What place is accorded the individual consumer in Japan's industrial policy priorities? The answer, it appears, is a place of secondary importance. Producer groups receive preferential treatment. During the decades of rapid growth, MITI officials concentrated single-mindedly on the promotion of producer interests.

Not that higher civil servants in MITI callously disregarded individual consumers. If queried, most MITI officials would probably have said that since producer and consumer interests are compatible, the growth of the Japanese economy would provide spillover benefits for individual consumers. "What's good for Japanese business," in other words, "is good for Japan as whole and therefore good for individual consumers."

Over the years, however, producer and consumer interests have not always converged. One can cite the following examples of specific conflicts: infant industry protection (for the telecommunications industry, for instance) versus more efficient production and service; protection of certain noncompetitive industries (like lumber processing) versus lower consumer prices; preferential treatment of farmers versus uncontrolled land prices; antirecession cartels (in the petrochemical industry) versus lower end-user prices; favorable tax treatment for small- and medium-sized enterprises versus lower relative tax burdens for individual households; and the power of small mom-and-pop stores to veto the establishment of large retail distribution outlets versus access to a broader and cheaper range of consumer products. In virtually all of these cases, the interests of producer groups have taken priority over those of individual consumers.

Certain characteristics of Japanese society and politics help to account for the dominance of producer interests. Consumer groups, like other horizontal organizations in Japan, have lacked the clout to champion consumer causes. Japan has no counterpart to such influential consumer advocates as Ralph Nader and John Gardner. And, as explained in greater detail in Chapter 3, MITI's institutional structure, centering on close, cooperative ties with industry, predisposes government officials to favor producer over consumer interests.

Since the early 1980's, however, three developments have weakened the primacy of producer interests: (1) the growing complexity and differentiation of producer interests; (2) the proliferation of ties of international interdependence; and (3) the enormous weight lent to Japanese consumers by foreign producers trying to break into the Japanese market. As outside pressures from foreign producers mount and strengthen the hands of domestic consumer groups, the pronounced bias in favor of producer groups is beginning to weaken. The government is being forced to pay more attention to consumer interests. But whether this trend will lead eventually to the enthronement of the consumer remains to be seen.

Skepticism about market capitalism extends beyond the concept of consumer sovereignty. It is probably most strongly felt on questions of industrial structure. Specifically, can the market, by itself, yield an industrial structure that meets national needs, not simply one that emerges passively out of the global division of labor?

Since, in the minds of MITI officials, the answer is clearly no, the government has no recourse except to turn to industrial policy. If the government takes no action, the market will produce an industrial structure that may not meet the nation's strategic aspirations and needs. Imagine how differently postwar history might have turned out if MITI had let the market take its course. Would a broadly based, heavy manufacturing infrastructure have emerged? If not, would Japan be in a position today to compete across the board in high technology?

Implicit here is the assumption that, under the right circumstances, the government is capable of utilizing its power to create the right mix of industries. The assumption comes close to the concept of dynamic competitive advantage, as distinct from the notion of static or passive comparative advantage. Instead of letting geography, national factor endowments, and the global division of labor dictate the shape of a country's industrial structure, the national government can determine where it holds competitive advantage by establishing a clear-cut set of priorities and administering a shrewd package of policies to cultivate priority industries. Factor endowments are not immutable legacies of Mother Nature. They can be changed through such industrial policies as manpower training, research and development, selective investments, and trade promotion.

Another problem associated with capitalist systems is the disequilibriating impact of business cycles. Japanese companies are especially vulnerable to downturns in the business cycle because of such distinctive features of Japan's industrial system as career-long employment, high debt-to-equity ratios, and the intensity of commercial competition. Owing to such characteristics of industrial organization, Japanese companies have a harder time taking belt-tightening measures that their foreign competitors routinely take: namely, laying off workers, cutting back production and investment, and operating at very low levels of plant capacity.

On the other hand, such rigidities are offset, to a significant extent, by the existence of other characteristics of Japanese industrial organization: namely, temporary workers, the system of subcontracting, close banking-business relations, ready access to capital, deep financial pockets to sustain extended periods of losses, and the outlet of export markets. Yet even with these offsetting mechanisms, Japan's industrial system is still vulnerable to the disruptive effects of sustained cyclical downturns, as evidenced by Japan's exceedingly high rate of corporate bankruptcies.

To cushion the downturns in business cycles, the Japanese government has been willing to organize or sanction antirecession and rationalization cartels, as temporary and strictly controlled measures.

It is hard to imagine a set of commercial circumstances under which the U.S. government would permit—indeed, initiate—the organization of cartels. In the United States, cartels are considered anathema to the country's economic well-being and system of democracy, a blight that must be avoided lest the vigor of market competition be squelched.

In Japan, the dangers of cartels are perfectly understood but the same ideological stigma is not attached. In fact, the Japanese believe that temporary cartels, used selectively and under the right circumstances, can prevent fratricidal warfare, high bankruptcy rates, and industrial disorder. In the short term, cartels can establish and maintain enough equilibrium to prevent a complete breakdown in industrial structure. Over the long run (assuming they can be phased out), cartels can prevent undesirable levels of market concentration. About cartels and business cycles more will be said later. Suffice it here to note the striking contrasts in views of cartels in Japan and the United States.

Excessive Competition

One of the most striking features of Japanese industrial policy during the era of double-digit growth, and one that is considered a distinctive feature of Japanese capitalism, is the phenomenon curiously referred to as "excessive competition" (*katō kyōsō*).[65] Excessive competition is often cited by MITI officials as the immediate reason for having to extend so visible a hand in the marketplace. On the positive side, it is sometimes given credit for supplying the kinetic energy behind the dynamism of Japan's private sector. On the other hand, excessive competition is said to have given rise to such problems as excess plant capacity, dumping and predatory pricing, and low company profits.[66] Whether cause or effect, the phenomenon has been a conspicuous feature of the Japanese political economy and needs to be understood, since it has legitimized the application of special industrial policy measures.[67]

What MITI officials seem to have in mind when they use the term *excessive competition* is a zero-sum situation in which an excess number of producers possess supply capacities that far exceed demand. Such disequilibrium gives rise to dysfunctional competition between firms, leading to severe price slashing, "forward" or "predatory" pricing,[68] a rash of bankruptcies, and industrial disorder. Since industrial chaos is harmful to the collective good, MITI officials feel they must restore order by containing the negative effects of excessive competition by applying industrial policy countermeasures.[69]

Left unchecked, the negative repercussions of excessive competition can be extremely troublesome. Excessive competition can lead to serious distortions in capital and labor allocations, opportunity costs with respect to potential scale economies of production, suboptimal levels of industrial concentration, a lack of export competitiveness, and loss of growth opportunities. Of more immediate concern, excessive competition is apt to generate or exacerbate problems caused by overinvestment in production facilities. Excess capacity is an especially serious problem in Japan because large Japanese corporations are saddled with high fixed costs as a result of permanent employment and heavy debt financing.

When recessions occurred in the 1950's and 1960's, the unacceptability of large layoffs or bankruptcies prompted the government to take action, organizing antirecession cartels to limit supply, restrain price competition, and agree on investment levels through administrative guidance.[70] For antirecession cartels to work, the influx of foreign goods simultaneously had to be curbed, because cheap imports would have undercut the effect of limiting supplies and establishing a floor for prices. To cope with the consequences of excessive competition, therefore, the Japanese government had to administer various market-bending countermeasures, transferring the costs of adjustments to Japanese consumers and would-be foreign exporters through a combination of antirecession cartels and import restrictions. Here, again, we see evidence of priority given to producer interests over the individual consumer.

For most capitalist states, the very idea of excessive competition sounds illogical, almost a contradiction in terms. Most mature capitalist economies, especially in Europe, face precisely the opposite problem: tepid competition and slack business investment demand. For them, the task is one of finding ways to keep competition and investment levels brisk. Why is it that Japan has faced so unusual a problem?

Industrial Policy as Root Cause

Various hypotheses have been advanced to explain the phenomenon of excessive competition. According to one hypothesis, the root cause can be traced directly to Japanese industrial policy.[71] By selecting certain industries for preferential development, and by cushioning the risks of business cycle downturns, Japanese industrial policy had the effect of stimulating overinvestment in plant capacity. Cynics might

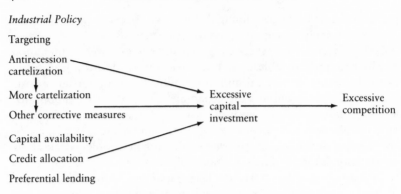

Fig. 1.1. Industrial policy as cause of excessive competition

even go so far as to intimate that MITI deliberately fostered excessive competition in order to consolidate its own power.

This line of argument is directly at odds with orthodox theories, which make just the opposite causal argument: namely, that excessive competition (cause) forced the government to adopt certain industrial policies in order to cope with the consequences (effect).[72] The sequence of causality is shown in Figure 1.1.

The industrial-policy-as-cause hypothesis does not clarify whether excessive competition was the product of calculated planning, or whether it came about inadvertently, as an unforeseen and unintended by-product of other industrial policy objectives.[73] But for proponents of this view, most notably Kōzō Yamamura, establishing conscious intent is not the issue. The point is that ill-advised market-bending industrial policies created conditions that bred excessive competition. To undo the damage it had done, MITI had to take countermeasures that embroiled it further in market manipulations.

Japanese industrial policy gave birth to excessive competition, the argument goes, by creating the incentives for companies to enter the fast-paced investment race.[74] Out of this reckless race sprang all the maladies associated with excessive competition—disequilibrium in supply and demand, vulnerability to cyclical downturns, antirecession cartels, possible violations of antitrust, overzealous price slashing, export deluges overseas, and trade frictions. If industrial policy had not skewed incentives to begin with, Japan would not have been afflicted with such severe problems. Proponents of this theory thus blame Japanese industrial policy for creating needless problems through the adoption of market-distorting measures.

This theory has much to recommend it. Certainly it exposes the simplistic nature of orthodox theories that explain MITI intervention on the grounds that the peculiar malady of excessive competition required unusual modes of market intervention. The industrial-policy-as-cause theory also has the merit of bringing out the complex interplay between industrial policy and microeconomic behavior. In making critical capital investment decisions, private corporations are bound to be affected by such industrial policies as special schedules for accelerated depreciation and preferential financing. Causality is usually interactive, not one-way.

In spite of its great virtue in calling attention to the interplay between MITI's policies and private-sector behavior, however, the industrial-policy-as-cause theory fails to provide a complete explanation of the origins and consequences of excessive competition. Although it might explain the phenomenon in, say, Japan's steel industry during the era of double-digit growth, it cannot account for its existence in such non-targeted industries as consumer electronics and precision equipment (such as watches and cameras). Nor, by the same token, can it explain why excessive competition has not occurred in some of the targeted industries, like aircraft, space, and lasers. If excessive competition is indeed the by-product of industrial policy targeting, why is it that some targeted sectors fail to exhibit the classic symptoms while other, non-targeted sectors show signs of suffering from the same affliction?

Since the dependent variable seems to be common, and does not vary perfectly with the independent variable, proponents of the theory would have to attribute its prevalence to systemic factors, or to "spillover" and "contagion effects." Or they would have to ascribe different causes to the existence of excessive competition in the non-targeted sectors. Either way, the theory cannot provide a complete explanation since factors other than industrial policy must be at work.

Japanese Management: Market Share Maximization Theory

To understand the etiology of excessive competition, we must consider alternative hypotheses that go beyond sector-specific explanations and incorporate causes more broadly embedded in Japan's political economy. One theory, advanced by specialists in Japanese management, ascribes the cause of excessive competition to the market share strategy adopted by Japanese corporations.[75] The aim of the strategy is to expand and hold significant shares of domestic and

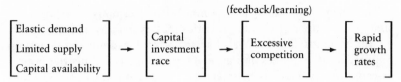

Fig. 1.2. Market-share maximization as cause of excessive competition

world markets, taking a long-term perspective on commercial competition.

This approach is distinguishable from the profit maximization strategy widely practiced in capitalist countries outside Japan, which concentrates on earning the highest returns on assets and investments on a quarterly or yearly basis. The route from market share maximization to excessive competition is shown in Figure 1.2.

When supply falls short of current or expected demand, companies rush to invest in new plant facilities, because adding new production capacity is the fastest way of expanding market share. Companies that fail to stay in the capital investment race—choosing instead to emphasize short-term profits—will find themselves falling farther and farther behind. Although profits may be buoyant for a time, market shares will shrivel up, and sooner or later these companies will find themselves squeezed out of the marketplace by corporations churning out products in mass volume at lower per-unit costs. Eventually, they will be forced to drop out of the competition. A market share maximization approach is therefore apt to accelerate an investment race, the sine qua non of corporate survival as well as the proximate cause of excessive competition.

For heavy investment to be sustained, demand has to be elastic and capital must be readily available. If either element is missing, the stampede to invest will not last long. The perception of inelastic demand is one of the main reasons excessive competition disappeared in all but a few industries following the two oil crises. Slumping demand dulled business incentives to invest, and capital investments plummeted after 1973. In the post-oil crisis recession, the behavior of Japanese firms in certain industries has come to resemble that of U.S. companies; more emphasis is being placed on earning high returns on investments.

This is not to say that Japanese corporations have discarded their emphasis on market share. To the contrary, a comparative study of Japanese and U.S. management, based on empirical data from over 500 corporations, has confirmed what individual case studies have long contended: that Japanese companies place market share at the top of their list of strategic priorities, clearly above such items as return on investment (ROI) and capital gains for stockholders. American companies, by comparison, rank these priorities in reverse order, placing return on investment and capital gains first and growth in market share only third.[76] Holding a share of the market, even in mature industries with flat demand curves, continues to rank at the top of Japanese corporate objectives. Only the rate of growth in capital investments has leveled off and with it have faded some of the most conspicuous features of excessive competition.

Why is it that an appreciation of the importance of market share is so widespread among Japanese but not U.S. firms? What is it about Japan that makes companies there disposed to place so much emphasis on market share? Peter Drucker attributes the cause to two features embedded in the structure of Japanese corporations: namely, lifetime employment and high debt-to-equity financing.[77] He argues that the imperatives of these two factors force Japanese companies to seek ever-expanding market shares. If Drucker is correct, Japan's extraordinary level of competition can be understood in terms of the convergence of three factors: high elasticity of demand (macroeconomic), lifetime employment, and high debt-to-equity financing (corporate management and structure).

To this list another crucial factor must be added: namely, corporate stockholding patterns and the willingness of institutional stockholders to accept low ROI.[78] Over 70 percent of outstanding Japanese shares are held by corporations and less than 30 percent by individuals. Corporate stockholders are willing to hang on to their stocks, even though the shares stay at par value and short-term dividends tend to be very modest at best, compared to U.S. corporations. Where Japanese stockholders receive compensation is in the actual (as opposed to par value) appreciation of their shares over time. By placing their bets on long-term stock appreciation rather than on short-term yields in high quarterly dividends, Japanese corporate stockholders can avoid regular payment of taxes on capital gains and corporate income and still reap higher long-term returns on invested capital.

Over the long run, if one looks at the bottom line, Japanese shareholders have come out ahead. They have gained more from stock appreciation than U.S. investors have earned from short-term dividends. According to one comparative study, from 1972 to 1982, Japanese stocks averaged better than a 14 percent annual return compounded, compared to only 7 percent by U.S. and West German stocks.[79] In addition, Japanese corporations derive tremendous flexibility and strength from intercorporate stockholding. It is a powerful deterrent to hostile outside takeovers[80] and a valuable means of consolidating ties and lowering transaction costs with business partners.

It is important to note that this pattern of stockholding means Japanese managers are under less compulsion to sustain high quarterly profits than their U.S. counterparts, and therefore freer to focus on long-term expansion of market share. Here, in short, is an underlying structural difference that helps to set Japanese capitalism apart from the forms of capitalism observable in the United States and many European countries (with the exception of West Germany). It places in clear perspective many of the fundamental distinctions than can be drawn between Japanese and U.S. institutions and behavior—differences in corporate strategy, industrial organization (especially the prominence of extra-market institutions in Japan), and industrial policy.

This structural distinction, however, may erode over time, as Japan's financial markets are increasingly internationalized, as individual shareholdings expand, and as capital-rich financial intermediaries, like pension fund managers and insurance companies, become stronger and more active on Japan's stock exchange. "This may imply," as Masahiko Aoki points out, "greater pressure on the management for share-price maximization."[81] Whether and how fast this happens is a matter of speculation; even if changes in shareholding patterns proceed rapidly, however, it is not clear that they will lead to a convergence in strategic objectives and patterns of corporate behavior, particularly with respect to share-price maximization.

The modified Drucker hypothesis has the virtue of linking microeconomic behavior to distinctive features of Japanese capital and labor markets. The argument is therefore anchored in structural variables that stay fairly constant over time. It does not have to rest on sector-specific industrial policies that are continually changing over time. It is a parsimonious theory. Special allowances do not have to be made for non-targeted sectors that confound theory by behaving like targeted sectors. But this virtue is also a weakness. For, though the Drucker hypothesis sheds light on broad, central tendencies, it requires modification or revision when deviations from the national mean are observable.

Not all Japanese companies practice permanent employment, for example. While firm figures are hard to find, estimates generally place the percentage of those covered by lifetime employment at somewhere between 25 and 35 percent of the labor force, making the practice considerably less than universal. Nor do all Japanese corporations operate on the basis of heavy debt financing. Such well-known giants as Matsushita and Toyota carry little or no debt. How can one explain the market share orientation of these companies? Is it because rival companies, which do pursue the growth of market share, force them to follow suit? Or do permanent employment and external indirect financing really compel Japanese companies to adopt a market share strategy, as Drucker asserts?

Falling Average Costs of Production Hypothesis

A third hypothesis, propounded primarily by Yasusuke Murakami, provides a slightly different interpretation.[82] The distinction between profit and market share strategies, Murakami argues, is not always clear-cut; the two are not necessarily mutually exclusive. Under certain circumstances—specifically, when the average, long-run costs of production are falling—the line of demarcation between the two is obliterated. The behavior of the firm out to maximize profits becomes indistinguishable from that of the firm out to capture the largest market share. Specifically, the way to increase profits—if average production costs are dropping—is by expanding output and market share.

If prices follow declining cost curves, it may be rational for profit-maximizing companies to stop somewhere short of optimal output. But as aggregate revenues increase with larger market share, even with marginal decreases in value added per unit, the incentives to expand output as much as possible remain strong. Of course, this picture is undoubtedly simplistic in that market prices probably would not stay in static equilibrium when long-run average costs are known to be falling. But precisely because the market is in a state of disequilibrium (in terms of price theory), companies seeking to raise profits are likely to aim at expanding market share. Indeed, such market instability, in Murakami's view, provides perhaps the only meaningful definition of excessive competition for the postwar Japanese economy (see Fig. 1.3).[83]

If this is so, then management theory positing the uniqueness of Japan's market share strategy (as opposed to the profit-seeking approach) does not explain the phenomenon of excessive competition; market share maximization is only spuriously related. The underlying cause is found in the declining long-run costs of production. Similarly, the industrial-policy-as-cause theory is found wanting: "If Japanese

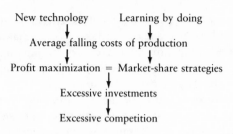

Fig. 1.3. Average falling costs of production as cause of excessive competition

industrial policy had not been put into effect as the believers of market mechanism would have advised, there would have still occurred a kind of investment race because of the potential of decreasing average costs."[84]

On the other hand, proponents of the first theory can argue that Japanese industrial policy may have helped to sustain the phenomenon of excessive competition, even if it did not bring the problem into being. If MITI had not cushioned the downside risks of aggressive investment, a rash of bankruptcies would have ensued, casting a pall over the enthusiasm for capital investment. Business leaders would have perceived long-run average cost curves as turning upward instead of downward, leading to a slowdown of the investment race and the elimination of excessive competition.

The declining average cost hypothesis is also attractive for its parsimony. Unlike the first two theories—MITI's targeted industrial policies and Japanese management strategy—it does not turn on particularistic features of Japan's political economy. The argument is universalistic. However, like the other two theories, it is not free of problems. There are technical questions, to begin with, about the degree to which profit and market share maximization actually converge. Moreover, if falling average cost curves lead to excessive competition, why has it not appeared as a problem in other countries? One would expect average production costs to have fallen elsewhere, if only because of the diffusion of innovative technology and the worldwide expansion of demand. If so, why have we not seen the spread of excessive competition outside Japan?

To deal with this puzzle, the cost-curve theory is forced to come up with variables that set Japan's capitalist economy apart from others, such as special characteristics of its capital markets. It has to fall back, in short, on exogenous factors that are unique to Japan. Thus, the falling cost-curve theory cannot be a necessary and sufficient explanation. This brings the various theories of excessive competition full

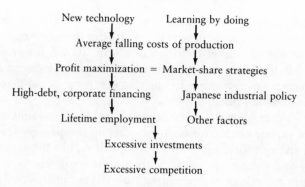

Fig. 1.4. Revised theory of average falling costs as cause of excessive competition

circle, with risk-reducing industrial policy and unusual aspects of Japanese management adding the final ingredients for a comprehensive explanation (see Fig. 1.4).

If the revised theory is accurate, it ought to be able to predict changes in microindustrial behavior over time and by individual sector. For example, one would expect Japanese companies' behavior in such sectors as steel, petrochemicals, and automobiles—where average costs are no longer steadily declining—to come increasingly to resemble that of their foreign competitors. By the same token, one would expect to see signs of excessive competition in those sectors where average costs are still falling—like the semiconductor, computer, and other high-tech industries.

Owing to marked differences in circumstances, however, the old mix of measures used to regulate excessive competition will probably be neither effective nor acceptable in terms of international norms of commercial behavior. Antirecession cartels, barriers against foreign direct investment, insulation of Japan's financial system, and all-out export campaigns—policies associated with excessive competition in old-line manufacturing sectors during the 1960's—have had to be sharply curtailed or abandoned altogether.[85] For the emerging high-technology sector, new forms of regulation have had to be devised—informal export restraints, mediation of trade conflicts, administrative agreements with the United States to increase foreign imports of high-technology products, facilitation of direct foreign investment both inside Japan and abroad, and so forth.

Even with the shift from smokestack industries to high technology, the need for government intervention as a means of compensating for market imperfections has not disappeared. Only the form and scope of intervention have changed. Y. Ojimi, former vice-minister of MITI, has called attention to the inadequacy of relying solely on the price mechanism in "new areas in which it is difficult to absorb enough risk capital by way of market distribution of capital, and the area of intangible goods, such as the research and development industry and the information industry, where the mechanism for rational determination of commodity prices is weak."[86]

Adaptive changes in the private sector—such as the establishment of a venture capital market for high-risk start-up companies—can serve to offset some of these problems; but for others, like the sponsorship of costly and risky basic research, the public sector seems to be the only entity capable of compensating for market imperfections. Japanese industrial policy is thus designed to cope with the positive and negative consequences of fierce competition in high technology, drawing on policy instruments notably different from those employed in the past for old-line industries. More will be said concerning these policy measures in subsequent chapters.

Market and State

The Market: Japan and the United States

Notwithstanding images of "Japan, Inc." or of a centrally planned economy, the Japanese government is actually closer to the United States and West Germany in its fundamental reliance on the market-place than it is to France, Italy, Britain, or certainly the socialist states. Like its U.S. and West German counterparts, the Japanese government considers the market mechanism to be the main engine of economic growth and industrial development. It believes that the market system is superior to any other ever devised. Among the market system's many virtues, it has demonstrated an unsurpassed capacity to stimulate technological innovation, raise productivity, motivate purposeful behavior and hard work, and elevate the material well-being of citizens. And quite apart from its economic merits, the capitalist system also has created the socioeconomic infrastructure necessary for complex, modern democracies to function on a stable, ongoing basis.

In light of such strengths, Japanese government officials believe that the market should be given as much leeway as possible to function, unencumbered by the long and sometimes stifling arm of the state. They are aware of the potentially heavy costs of unwarranted government intervention. When the government is unable to resist the temptation to tamper, the dynamism and efficiency for which the market system is known can easily be undermined. On the basis of the capitalist creed—namely, reliance on the market to function as the driving engine of growth—the Japanese and U.S. governments therefore have much in common.

Where the Japanese and U.S. governments part company is in the scope and specificity of goals set forth and in the perceptions of the market's capacity to achieve these goals. Simply put, the Japanese government sets a longer, more concrete, and more ambitious agenda of goals that it feels the self-regulating market cannot be counted on to achieve. The U.S. government is much less specific and ambitious and more confident in the market's capacity to produce favorable outcomes, even if the precise nature of such outcomes cannot always be predicted. The Japanese leave less to chance.

The Japanese approach to the market bears some similarities to their aesthetic view of bonsai and traditional Japanese gardens: one must work within, and respect, the essential forces of nature; but one must enhance nature by utilizing man's unique capacity to adapt, structure, and highlight the elements of nature in ways that produce a blend of natural and man-made beauty. Man can design the overall composition of a bonsai or garden, bend trees into all sorts of exquisite shapes, and produce a work of art that is consciously designed but still completely natural. Instead of letting nature take its course, man can actually improve on what nature produces.

The bonsai analogy is reflected in the existence of deep-seated doubts that the unfettered market can advance the common good and achieve specific national goals. It will not, for example, generate the large-scale investments needed for basic research in high technology; nor will it guarantee the nation's economic security, or overcome the disadvantages of being behind. To accomplish these goals, the government must take the initiative and engage in some economic engineering, within the general framework of market forces.

MITI thus sets specific objectives that it wishes to achieve. During the remaining years of the twentieth century, these objectives include: (1) shifting Japan's industrial structure from energy-intensive heavy manufacturing to knowledge-intensive high technology; (2) creating a stable and supportive business environment; (3) reaching the state-of-the-art frontiers in high-technology research and development; (4) improving economic efficiency and productivity; (5) improving the quality of life; (6) ensuring economic security; (7) integrating Japan's industrial economy smoothly into the international economic system. In contrast, the U.S. government seldom defines a clear set of goals for individual industries or sectors. It usually only responds to problems as they arise.

Not that the U.S. government is willing to sit back and see its industrial economy go wherever the self-regulating market takes it. Jobs have to be maintained, key declining industries kept alive, and the demands of politically powerful special interests satisfied. Indeed, in certain areas, like the correction of social injustices (such as job discrimination), the U.S. government is far more interventionist than its Japanese counterpart. The difference is that the Japanese confine themselves, by and large, to economic, not social, goals, and the policy measures used to reach those goals tend to fall into the category of what Assar Lindbeck calls "market-conforming methods of intervention,"[87] not "market-defying" methods, which usually breed inefficiency.

Intervention Based on Industrial Life Cycle

MITI's "market-conforming methods of intervention" are nowhere more clearly seen than in its use of the concept of industrial life cycle as a basis for determining what might be considered an appropriate level and form of state intervention. Although the degree of intervention and selection of policy instruments vary by industry, MITI intervention tends to follow a curvilinear trajectory: that is, extensive involvement during the early stages of an industry's life cycle when market demand is still small, falling off significantly as the industry reaches full maturity and demand reaches its peak, and rising again as industry loses comparative advantage and faces the problems of senescence—saturated markets, the loss of market share, and excess capacity.

Figure 1.5 illustrates the relationship between government intervention and the stages of the industrial life cycle. The solid line reflects the level of growth in demand at each stage of industrial development. A number of high-technology industries are clustered at the early stage—data processing, computer services, biotechnology, space, ocean development, and nuclear energy. Computer hardware and aircraft are located further up the trajectory, with semiconductors (integrated circuits) near the apex of development. The dotted line indicates the level of state intervention at each stage of development, with the highest levels coming at the early and late phases of the industrial life cycle.

Japanese industrial policy for the semiconductor industry is an example of early involvement followed by some degree of disengagement.[88] Because Japanese producers lag behind U.S. producers in

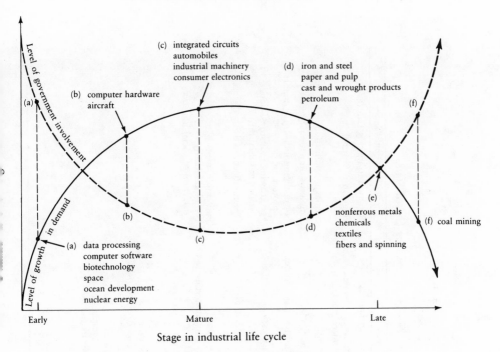

Fig. 1.5. Industrial life cycles and government intervention

the most sophisticated areas of semiconductor technology, such as logic chips, and because of the public-goods nature of basic, precommercial research, the government has not disengaged completely. It still has its hands in R&D activities. But it has jettisoned most infant industry measures, opened up procurements in telecommunications, and curbed export dumping practices. MITI is giving the semiconductor industry freer rein than it did in the early 1970's. The disengagement is in keeping with MITI's assessment that greater intervention would serve no useful purpose. What the state can do once an industry has matured is limited.

A United States–Japan Comparison

Tables 1.2 and 1.3 outline the major commonalities and differences between Japan and the United States with respect to the circumstances that give rise to government intervention. As indicated in the outline, the Japanese government views its role in microindustrial management as far broader in scope than its U.S. counterpart, especially in the two areas of market imperfections and industrial catch-up. Only in

TABLE 1.2

Government Rationales for Intervention, Japan and the United States

Japan and the United States	Japan	United States
Market failures	Anticipation of market failures	Reaction to market failures
Externalities	Administrative guidance; private sector compliance; regulation	Regulatory control
Antitrust	Flexible enforcement	Stringent application
Public procurement	Emphasis on public works	Emphasis on national security
Adjustment to shifting international comparative advantage	For declining industries: reduction in plant capacity For growing industries: competitive advantage	For declining industries: import protection For growing industries: hands-off stance or initial thrust
Regional dispersion of jobs and industrial location	Centrally coordinated: MITI	Decentralized: individual states
National priorities	Industrial catch-up (past) Economic security (raw materials and overseas markets) Consensus formation; broad future directions Industrial orderliness (fears of excessive competition) Equitable distribution of costs, risks, and benefits Coping with fluctuations in business cycles and exchange rates (especially for small and medium-scale companies) Industrial structure (based on ever-higher value added) Improved quality of life	Full employment National security Correction of social injustices

TABLE I.3

*Conceptual Framework for Government Intervention,
Japan and the United States*

Japan	United States
Market imperfections	*Market failures*
Capital market deficiencies	Externalities
Excessive competition	Neglect of collective good
Regional maldistribution of resources	Antitrust abuse
Industrial disorderliness	Business cycles
Production inefficiencies	Manpower needs
Resource misallocations (nonpriority	Excessive risks
sectors)	Unemployment
Problems related to industrial structure	Redistribution
	Social injustices (need for affirmative
	action, etc.)
	Loss of international competitiveness
Economic security	*National security*
Structural maladjustments	Supply disruptions (raw materials)
	Foreign market closure
	Dangerous foreign dependence
	Loss of competitiveness in vital
	industries
	Need for technological edge
Industrial policy	*Distortions from government*
fallout effects	*intervention*
Assistance for small and medium-sized	Contagion effect of policies (taxes,
enterprises	subsidies)
	Remedial policies
Industrial catch-up	
Infant industry vulnerabilities	
Threat of lower value added;	
unacceptability of certain areas	
of comparative advantage	
Loss of industrial autonomy	

national security, antitrust, and the protection of declining industries is the United States more active. But the costs of state intervention based on the primacy of political considerations can be high, since it is usually market-defying.

The essence of Japanese industrial policy can thus be summarized as a strong orientation toward development, world competitiveness, efficiency, flexible adaptation, and the pursuit of national interests. The complex reasons that Japan, of all the major states utilizing industrial policy, stands out as perhaps the most successful example can be attributed largely to the strengths of its private sector and to its political system (which will be analyzed in greater detail later).

But the conceptual framework used to formulate and administer industrial policy is not an insignificant factor. Of particular importance has been MITI's general adherence to market-conforming methods of government intervention (with some noteworthy exceptions), a philosophical tenet that has kept the state from succumbing to the temptations of political expediency. This has largely spared Japan the fate that has befallen most countries that have relied heavily on industrial policy: namely, costly subsidies to inefficient and noncompetitive industries. When subsidies have been offered, they have reinforced, not stifled, incentives to upgrade efficiency and the capacity to compete in world markets. In most cases of industrial policy elsewhere, the opposite has been true. MITI's capacity to resist political quick fixes is a key to understanding what is otherwise a paradox: namely, an active government, intervening broadly in the market, yet comparatively small in terms of its claim on GNP and possession of formal legal authority.

TWO

Industrial Policy Instruments for
High Technology

Going into the last quarter of the twentieth century, U.S. companies completely dominated the so-called high-tech fields, just as U.S. producers had once dominated the old-line manufacturing sectors. The United States, postwar birthplace for nearly all high-tech industries, held a commanding lead in such areas as semiconductors, computers, aerospace, telecommunications, new materials, and biotechnology. Most of the innovative breakthroughs that gave birth to glamorous new industries were made in the United States: the integrated circuit, vacuum-tube computer, microprocessor, digital switching equipment, fiber optics, nuclear fission, and recombinant DNA, among others. Even today, most of the leading corporations in these fields are American: IBM (semiconductors and computers), AT&T (telecommunications), Boeing and McDonnell-Douglas (commercial jet aircraft), and Lockheed, Martin Marietta, Rockwell, and Hughes (aerospace).

Not long ago, the same could be said of the old-line manufacturing sectors. From the end of the Second World War until the late 1960's, U.S. producers of automobiles, steel, and color televisions led the world in market share, technology, and profitability. Both the car powered by an internal combustion engine and the color television set were invented in the United States, and the names of leading U.S. corporations were almost synonymous with the product itself: General Motors and Ford (automobiles), U.S. Steel, and RCA and Zenith (color televisions).

By 1970, however, Japanese latecomers had caught up with and overtaken U.S. front-runners in steel, automobiles, and consumer electronics. The speed with which the Japanese closed the gap and opened up a widening lead of their own startled everyone, including the Japanese themselves. Japan's capacity to overcome what had seemed, at one time, like an insurmountable disadvantage has been attributed, at least in part, to the momentum imparted by Japanese industrial policy.[1] How else could a country come from so far behind so swiftly?

For years, even while U.S. producers of cars, steel, and consumer electronics were receiving a sound thrashing in their own marketplace, U.S. front-runners in high technology could take comfort in the thought that their industries were not as vulnerable as those in the smokestack sectors. Unlike steel and automobiles, which played to Japan's strengths in mass production and standardized process technology, competitiveness in semiconductors, computers, and other high-tech industries depended on new product innovation, state-of-the-art design, creative software applications, and complex systems integration—areas in which the Japanese were thought to be weak. The pivotal importance of innovation thus lulled some Americans into a false sense of security. Semiconductors and computers seemed safe from Japan's commercial offensive; high technology was the last bastion of U.S. comparative advantage.

The sense of security did not last long. By 1988, leading U.S. corporations found themselves embroiled in a raging war against Japanese corporations on virtually all fronts of high technology, from lasers to pharmaceuticals. The battle in semiconductors seemed especially fierce and decisive in the eyes of the combatants. It was a battle Americans seemed to be on the verge of losing. By 1986, at any rate, Japan had surpassed the United States to become the world's largest producer of semiconductors, with 45.5 percent of world market share, compared to the United States' 44 percent; only eight years earlier, in 1978, U.S. companies had held 56 percent of the world market whereas Japanese firms accounted for only 28 percent.[2] Of the world's top ten semiconductor companies, six, including the top three, were Japanese. In certain standard, high-volume products, like dynamic random access memory chips (DRAMs), the Japanese had come to command 90 percent of world market share. Even in the technologically more complex products, such as microprocessors, microcontrollers, and applications specific integrated circuits (ASICs), the Japanese were making swift and significant enough inroads to set off warning sirens (if not Cassandra-like prophecies of doom) in Silicon Valley.

How has Japan done it? How has it managed to confound skeptics in fields where it was thought to be weak? In the minds of many non-Japanese, the answer resides in the effectiveness of MITI's industrial policies for high technology.[3] Specifically, MITI is believed to have given the Japanese semiconductor industry a powerful thrust forward by applying some of the same instruments of industrial policy that had been used for the old-line sectors: namely, industrial targeting, infant industry protection, controls over foreign direct investment, close government-business collaboration, access to cheap capital, encouragement of excessive competition, R&D subsidies, buy-Japanese programs, dumping and predatory pricing in overseas markets, and other unfair trade practices. Against the advantages conferred by industrial policy targeting, the small, innovative, and market-driven firms in Silicon Valley, operating without the benefit of government support and protection, appeared hopelessly overmatched. Like steel and auto producers, many felt they could not compete against "Japan, Inc."

The view summarized above, however, gives Japanese industrial policy more credit (or, depending on one's point of view, more blame) for the development of the semiconductor industry than it deserves.[4] Although it did play an indispensable role in the emergence of the steel industry (as pointed out in Chapter 1), Japanese industrial policy has contributed less to the success of high-tech industries.[5] MITI has applied a less extensive set of policy instruments, one more attuned to the different functional needs of high-technology endeavors. Recognizing that a world-class semiconductor industry could emerge only from the crucible of market competition and that, for the long-term health of the industry, the government should hold interventionist impulses in check, MITI refrained from using several strong instruments that had been applied to accelerate the growth of the steel industry: investment guidance, production targets, antirecession and rationalization cartels, extensive involvement in upstream activities, stringent standards of approval for technology transfers, strict controls over foreign direct investment, and mergers to achieve economies of scale. MITI did try to promote mergers in the computer industry, but in the face of private-sector resistance the idea died an early death.

In choosing policy instruments to fit the needs of specific high-tech industries, MITI has demonstrated a capacity to deftly adapt industrial policy to varying circumstances and to resist the temptation to intervene beyond the point of usefulness. It realizes that overzealous attempts to wean an industry can create perverse incentives, which can undermine rather than strengthen an industry's competitiveness. The combination of restraint, selective intervention, and respect for the discipline of market forces has made for an effective formula. This chapter will examine the concrete policy measures that MITI chose to administer, given the functional needs of high technology and the range of policy instruments available. Let us begin with an analysis of the functional requisites.

Functional Requisites

Industrial Dynamics

Although high technology differs fundamentally from smokestack industry, the two sectors share some characteristics. In both cases, manufacturing is a key ingredient of latecomer catch-up and commercial success. In both cases, attention needs to be paid to the upgrading of process technology; this, in turn, ensures that product quality will be continually improved. For standard high-tech products, as for most products of heavy manufacturing, the longer the production run, the lower the per-unit costs of production. The advantages gained from volume production suggest that it is vital to take a global view of markets and to export aggressively, especially to large overseas markets. Concerning such basics, high technology and heavy manufacturing have much in common.

The two differ, however, in the scope of technological change, the uncertainties, risks, and costs of research and development, the length of product life cycles, the complex relationship to downstream applications and end-user demand, and the structure of industrial organization. On each of these dimensions, the difference is one of degree, not of kind; but the amplitude of variation is so wide that the differences in degree add up to a difference in kind. At least the differences require the administration of distinct policy instruments.

None of the differences is more striking than the potential for technological change. The opportunities for technological advancement are much greater in high technology, because know-how is at an earlier stage of maturation. In older industries like steel, the chances of developing revolutionary new products—products that transform the very nature of competition—are exceedingly low. In the semiconductor industry, by contrast, major new products, like the microprocessor, have emerged at fairly regular intervals to alter the basic direction of industry, and no end is in sight. This gives small, entrepreneurial companies opportunities to find and exploit niche markets that are constantly opening up. In heavy manufacturing, pursuing a strategy of product differentiation is harder because there is less leeway to bring new products to market.

Because products, manufacturing processes, and competitive dynamics change constantly, companies must sink substantial resources

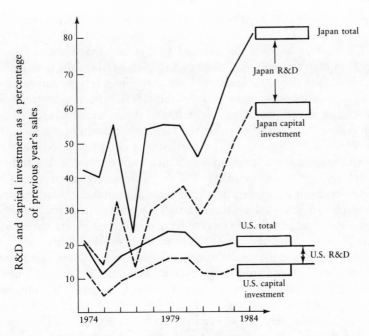

Fig. 2.1. R&D and capital investment in the semiconductor industry: United States and Japan, 1974–1984. Source: Dr. Yoshio Nishi.

into research and development. Over the years, Japanese semiconductor companies have ploughed back about 20 percent of their revenues into R&D (Fig. 2.1), an extraordinarily high rate compared to most old-line industries, which typically invest 3 percent or less. If companies fail to sustain high investment levels, they run the risk of falling hopelessly behind and of being forced out of the race altogether. To survive in the marketplace, high-technology companies must stretch themselves to the limit in R&D resource allocations. The level of R&D spending is, in fact, one way of defining the difference between old-line and high-technology industries.

Although the functional imperative of sustaining high levels of R&D investment is clear, the risks and uncertainties of R&D are substantial, and the investment of large sums by itself is no guarantee of success. The ratio of cost to commercial yield rises as companies strive to push beyond the frontiers of knowledge. Over time, the uncertainties and the rising ratio of costs to commercialization can strain the financial resources of competing corporations, since they must also simultaneously sink even larger sums into new plant facilities (see Fig. 2.1).

The cost problem is exacerbated by the fact that state-of-the-art R&D in high technology is close to and even overlaps the frontiers of knowledge in the basic sciences. Most companies feel they must have a firm foundation at both ends of the R&D process—upstream in basic science and prototype research, and downstream in applied development. The problem is that the farther one pushes upstream, the more research takes on the properties of a collective good; it is an inexhaustible collective resource to which everyone has access and for which no one is willing to pay.[6] Since upstream research has to be done—without it, applied development will eventually bump against the limits of knowledge—the question of who pays is problematic. Corporate budgets already tend to be overstretched by investments in new plant facilities and downstream product development.

For companies, the uncertainties and costs of R&D are aggravated by the shortening of product life cycles. Even when R&D investments yield commercial products, the period of time during which companies can capture rents and recoup the upfront costs of R&D has become shorter and shorter. Unlike steel products, which may sell for more than a decade, or car models, which may sell (with slight modifications) for several years, the typical life cycle for a DRAM might last less than a year and a half. The brevity of the product life cycle thus increases the costs and risks of R&D. The problem of sustaining heavy capital investments is especially acute during cyclical downturns in business demand.

The character of high technology also gives rise to diverse and complicated industrial structures. In the United States, where the semiconductor industry began, a multilayered structure of firms has emerged, consisting of giant captive (in-house) corporations, like IBM and AT&T, with full downstream capabilities, diversified merchant companies (which sell semiconductor devices to other companies), like Motorola and Texas Instruments, with considerable system-design expertise, nondiversified merchant producers, like Intel, Advanced Micro Devices, and National Semiconductor, with little or no downstream capability, and small, specialized niche producers, like LSI Logic, VLSI Technology, and Cypress, that concentrate on fairly narrow niche markets.[7] Rounding out the U.S. structure are a group of small, independent semiconductor equipment makers, which supply vital know-how in process technology, and large military houses, which purchase semiconductor components for installation in sophisticated weapons systems. Overall, the multitiered structure has met the functional imperatives of innovation in the growing semiconductor industry very well, though questions have been raised about the continuing viability of the structure as it is presently constituted.

The structure of the Japanese semiconductor industry poses a sharp contrast. The largest producers are big, vertically integrated, systems-oriented corporations, like NEC and Toshiba, that manufacture for in-house use and for external sales and also buy from rival producers. Smaller consumer electronics companies, like Sanyo and Sharp, similarly manufacture for in-house use and external sales and also purchase from the outside market. The stratum of small merchant producers, which has contributed so much to new product design and innovation in the United States, does not exist in Japan. There are many small semiconductor equipment manufacturers, but the most prominent of them are closely tied to the large systems houses through equity holdings and other ties of interdependence. Just as the heterogeneous industrial structure has served the needs of the U.S. semiconductor industry, so too the less heterogeneous, more vertically integrated structure has met Japan's needs as a latecomer, especially with respect to technological catch-up and worldwide market share expansion.

As the semiconductor industry reaches technological and commercial maturity, and as fierce competition continues unabated, the advantages and disadvantages of each industrial structure will become more apparent and certain adjustments may have to be made. In the United States, the mid-sized merchant houses, lacking the deep pockets, diversified product portfolios, and extensive marketing and service networks of the large Japanese corporations (and of giant U.S. captives), may feel most acutely the squeeze of competitive pressures. All U.S. companies will feel the effects of a decline in the market position of the small equipment manufacturers, especially in the area of process technology. Thus, the long-term trends may be moving in the direction of industrial restructuring, greater market concentration, vertical integration (though there will be room for small niche producers), and downstream diversification.

For Japan, the challenge ahead will center on the need to be innovative at the frontiers of both new product design and process technology. To date, Japan's structure of large, vertically integrated corporations has not lent itself to new product innovation as readily as America's multilayered structure. Even though size can mean deeper pockets for R&D investment, it frequently gets in the way of research flexibility and quick turnaround. One structural adaptation in Japan has been the "hiving off" of R&D to small, subsidiary companies and subcontractors, a pattern that Masahiko Aoki calls, in a broader context, "quasi-disintegration."[8]

Another structural response, involving both Japanese and U.S. companies, has been the proliferation of international alliances, such as joint ventures, mergers, acquisitions of foreign firms, venture capital investments overseas, original equipment manufacturing, and cross-licensing agreements. Strategic alliances have sprung up everywhere, reflecting the severity of global competition. There is no sign yet that the proliferation will come to a halt. So long as the functional requisites of high technology remain demanding, the internationalization will continue.

For Japanese and U.S. companies alike, the functional requirements of high technology are challenging and clear. They revolve around the need to advance technology beyond the frontiers of knowledge. To do so requires access to financial and human resources, sustained investments, simultaneous pursuit of new product design and process technology, the assumption of high risks and uncertainties, sufficient production experience to move along steep learning curves, the formation of international strategic alliances, and ongoing adaptations in industrial structure.

Government Requisites

In order to facilitate efforts by the private sector to meet these challenging functional requirements, national governments must do what they can to create a supportive business environment. Thus, in parallel with, and closely related to, private-sector requirements, there are government requisites that must be met if countries want to maximize their chances of achieving world competitiveness in high technology.

As most high-technology products have high demand elasticity, macroeconomic policy measures aimed at generating greater aggregate demand are bound to quicken the growth rate of high-technology industries. Similarly, low inflation levels, stable exchange rates, and sound fiscal management not only are vital ingredients of a healthy economy but also serve to cushion downturns in business cycles. Because investment levels in high-technology industries are sensitive to cyclical movements, anything the government can do to flatten out the dips in business demand will help companies maintain high levels of capital investment. Thus, sound macroeconomic policies are essential to the creation of healthy high-tech sectors.

Leaving aside macroeconomic measures, the government must also overcome the collective goods problem associated with basic research. If no one is willing to assume the costs of basic research, and progress in high technology hinges on breakthroughs in basic knowledge, then the government must find ways of supporting basic R&D. It can either allocate public funds or offer incentives for private-sector investment.

The government can facilitate the development of commercial R&D by offering tax incentives, organizing cooperative projects, furthering research economies of scale, relaxing antitrust provisions, eliminating needless duplication, and transferring technology from government laboratories to the private sector. Moreover, in Japan, the system of career-long employment tends to erect de facto barriers to the diffusion of technology between firms, a problem that does not beset more mobile labor markets, like the United States', where the movement of workers from company to company serves to diffuse know-how. From an economic point of view, the diffusion of technology is at least as important as product innovation, for it is only through the processes of diffusion that economic growth and commercial development take place.

High-technology industries are also dependent on the quality and quantity of technical human resources available. Here again, the burden of assuring that the human resource needs are met falls, in part, on the government's shoulders. The government's responsibilities include public education and vocational training, though the private sector can be expected to play a role as well. The Ministry of Education is in charge of Japan's educational system; and MITI, since it is not involved in curricular planning, has not been able to adjust the distribution of university students in various disciplines to match human resource needs within industry. Nor has it had much input in discussions of how to go about reforming the educational system in ways that bring out greater creativity in Japanese students (in keeping with the national need for innovation in high technology). MITI is involved in certain areas of vocational training, such as the establishment of minimum standards for certification in software programming.

Where MITI's industrial policy has had an impact is on the employment preferences of college graduates. When MITI promulgates its long-term "vision" for Japan's industrial future, Japanese graduates can identify the high-growth industries of the future. With this knowledge, choosing what industries and companies to enter can be easy. Over time, the best and brightest from Japan's finest institutions of higher learning have tended to gravitate en masse to jobs in rapidly growing industries: steel until the late 1960's, automobiles until the mid-1970's, electronics and the information industries since the mid-1970's.

In the United States, by contrast, the best and brightest tend to scatter in all directions. The most capable graduates appear to make more individualistic, short-term decisions (owing, in part, to the very mobile labor market). There is less responsiveness to industry-specific trajectories. It would be risky to forecast, for example, that there will be a swelling tide of graduates entering high technology; such traditional fields as law and finance, offering big bucks to young professionals, continue to skim more than their share of the cream off the United States' college crop. In neither the United States nor Japan does the government make a concerted effort at steering students directly into fields of study on the basis of the anticipated needs of the economy.

An area in which the government is expected to play an active role, however, is international trade. Since trade in high technology can engender serious conflicts, the government must try to resolve trade disputes that arise between domestic producers and foreign competitors. This task can take many forms: bilateral and multilateral negotiations, trade agreements, enforcement of fair rules and procedures, information gathering, and adjudication of grievances.

In all the above areas—demand stimulation, technology push, and trade mediation—the Japanese government has felt an especially strong sense of responsibility, owing to the nature of the country's finely meshed industrial system, which functions on the basis of competition and cooperation, market and hierarchy, public- and private-sector coordination, structural interdependence, "no-exit" relationships, and integrative consensus. Without government support, the high-technology industries in Japan would have a harder time meeting the demanding functional requirements of commercial success in competitive world markets.

Industrial Targeting

U.S. industrial and government leaders have voiced strong objections to MITI's practice of what they call industrial targeting: that is, the identification of strategic industries and their promotion through preferential treatment, funneling of large subsidies, infant industry protection, export promotion, and buy-Japanese programs.[9] The effect of MITI targeting, according to some U.S. critics, is that it creates a playing field tilted unfairly in Japan's favor, since the U.S. government does not engage in similar interventionist practices. High-tech companies in the United States are left to fend for themselves.

Although Japanese government officials dislike the term *targeting* because it carries connotations of unfair state intervention, they do not deny that they select certain industries for preferential promotion. The U.S. International Trade Commission (USITC) defines industrial targeting as "coordinated government actions taken to direct productive resources to help domestic producers in selected industries become more competitive."[10] A number of countries around the world engage in the practice, so defined, but with varying degrees of success. To engage in targeting is no guarantee of competitive advantage. Even Japan, regarded as the most successful, has made its share of mistakes, pointed out in the last chapter.

In semiconductors, for example, MITI has not always exercised good judgment. In 1953, when a small company called Tōkyō Tsūshin Kōgyō sought permission to purchase Western Electric's transistor technology for $25,000, MITI was reluctant to grant approval, citing a shortage of foreign currency. The use of scarce foreign currency for a technology with uncertain commercial applicability appeared to be a poor risk, particularly since the company making the application was a small start-up without much of a track record.

Only after a contract had been signed with Western Electric did MITI, in 1954, authorize the transfer of transistor technology.[11] That small start-up, subsequently renamed the Sony Corporation, went on to revolutionize the field of consumer electronics by successfully installing the transistor in small, portable radios and television sets. Thus began the process of miniaturization, which caught the fancy of consumers with its low cost and the convenience of compactness.

Imagine the enormous opportunity costs to the Japanese electronics industry if Tōkyō Tsūshin Kōgyō had not gone ahead to sign the transistor patent agreement, which it presented to MITI as a fait accompli. The episode belies the myth of MITI's prescience. It also brings to light the fact that some of Japan's most successful export industries—consumer electronics, cameras, watches, and other precision equipment—have managed to grow up strong and healthy outside MITI's incubator for targeted infant industries.

What about the charge of unfairness? Do the instruments of industrial targeting tilt the playing field unfairly in Japan's favor? To answer this question requires an in-depth examination of Japanese industrial policies for high technology, comparing them wherever possible to those of the United States and Western Europe. The miscellaneous measures will be grouped under three broad categories: technology push, other facilitating measures, and demand pull. An effort will be made to assess the relative importance of these policy instruments with respect to the development of Japan's high-technology sector.

Technology Push

National R&D Systems

Before analyzing the concrete measures taken to accelerate the pace of technological change, it might be useful to compare briefly, as background, the R&D systems of the United States and Japan. Probably the biggest differences lie in the scope of military R&D, the proportion of government expenditures, and the links between universities, industry, and government.[12] Japan has the luxury—enjoyed by no other large industrial economy—of being able to concentrate almost exclusively on commercial R&D. By contrast, the United States bears the onerous responsibility of providing for the common defense of the Western alliance. The ramifications of this disparity in defense burdens are far-reaching.

Japan's purely commercial orientation is a major asset because it permits the government to make cost-effective use of the country's R&D resources. Few of its scientists and engineers, and only a small portion of its budgetary resources, have to be diverted to military R&D projects. Since the commercial spillover from narrowly specialized military R&D projects tends to be limited, the diversion of finite resources means that substantial opportunity costs are incurred by the United States, France, and other countries spending large amounts for military purposes.

To be sure, the U.S. military-oriented R&D system has served as the womb of nearly all high-technology industries during the postwar period. The U.S. government's role as R&D contractor and guaranteed first customer has fostered the growth of one high-tech industry after another. Following infant nurturance, each industrial offspring has reached maturity and independence with the expansion of commercial demand in the U.S. market, the largest in the world.[13] The U.S. R&D system has thus functioned like no other in the world to create whole new industries of enormous economic and technological significance—semiconductors, computers, supercomputers, telecommunications, and many others.

The same R&D system, however, so admirably suited to the creation and early development of high-technology industries, has had trouble coping with the latecomer challenge posed by Japan. This is because there appear to be hard trade-offs between military and commercial R&D. Military and commercial technologies, which overlap during gestation and the early years, tend to separate as fledgling industries mature. Zero-sum investment choices have to be made.

The inefficiencies, rigidities, and commercial opportunity costs built into the U.S. military R&D system may render the overall R&D system less adaptable than Japan's, at least in the commercial realm. If the strength of the U.S. system lies in its pioneering character, the strength of Japan's system lies in its capacity to convert breakthroughs in basic knowledge swiftly into tangible products on store shelves. Japan's commercially oriented and market-driven R&D system is based on painstaking consensus building, government-industry cooperation, an emphasis on advancing the not very glamorous but commercially decisive area of process technology, cost-effective resource allocation, information sharing, technology diffusion, and a singularity of focus on commercial applications.

Until the 1980's, the Japanese R&D system was geared to latecomer catch-up, and the above-mentioned features contributed to Japan's remarkable capacity to close the gap with front-runner countries. However, as Japan reaches the frontiers, it is striving hard to adapt its latecomer system to one better suited to pioneering breakthroughs, while retaining its strengths in process technology. The shift from latecomer to pioneer will not be easy. To innovate at the frontiers, Japan must overcome some glaring deficiencies, such as the lack of synergistic interaction between universities, corporations, and the government. This will require Japan to upgrade its upstream activities in basic scientific research and precommercial development. This, indeed, is what has prompted the Japanese government to organize an ambitious series of national research projects aimed at pushing Japan beyond the frontiers of technology.

National Research Projects

From the standpoint of cost-effectiveness, national research projects gathering together talent from the leading companies and government laboratories represent an ideal way of leapfrogging ahead. The trouble is that few countries are capable of organizing them effectively. Either the companies are not sufficiently competitive, or the distrust between them is too deep-seated to be overcome; more often than not, the disincentives outweigh the incentives to cooperate, and sometimes antitrust is too strictly enforced.

Japan is one of the few countries in the world that has demonstrated that it can organize national projects successfully. The reasons are too complex to go into here, but perhaps the concentration of R&D talent in a manageable number of leading firms and a relationship of trust between government and industry can be cited as noteworthy reasons.[14] National projects for the information industries began in earnest in the early 1970's; the 3.75 Series Computer Development Project (1972–76) was Japan's response to IBM's development of a third-generation computer. A flurry of national projects followed in rapid succession, the best-known of which have been the very large scale integrated (VLSI) circuits and Fifth Generation Computer projects.

It is no coincidence that the number of R&D projects increased markedly at just the time when Japan was feeling the force of foreign pressures to liberalize its import tariffs and investment barriers. MITI officials and industry leaders, fearing that liberalization would expose the weaknesses of Japan's information industry, felt that the level of Japanese technology had to be advanced through a crash program of national R&D projects. The dismantling of tariff and investment barriers thus forced Japan to turn to national projects as a vehicle for technologically leapfrogging ahead. Fortunately for Japan, the information industry had already developed to a point where it was poised to make a great leap forward. Had it lagged further behind, national projects might not have been enough to advance it to its present level of competitiveness.

The focus of all national research projects in Japan is on basic, precommercial technologies of such seminal importance that interfirm cooperation makes eminent sense. The projects reveal several common characteristics: (1) the development of precommercial prototype products, (2) long gestation periods, (3) high uncertainties and risks, (4) heavy capital outlays, (5) research economies of scale, (6) steep learning curves, (7) the promise of advancements in process technology, and (8) potential commercial utility across an array of industries. They have to focus on upstream technologies that are so costly and risky that individual companies, left to their own calculations of cost-effectiveness, would probably not make the investment. The government must step in where market incentives alone might not be sufficient to promote the collective interests of priority industries and of the national economy. Projects must also hold out the promise of multiplier-effect benefits.

MITI feels a public and national responsibility to do whatever it can to give domestic producers a competitive edge over foreign companies. The organization of national projects provides tangible evidence that it is doing something of high visibility and strategic value to promote both Japanese industry and the public well-being. Such projects also offer MITI a concrete and effective mechanism for shoring up its power, consolidating ties with the private sector, and coaxing more money out of the Ministry of Finance. Such secondary effects should not be dismissed as incidental by-products of national research projects inspired by more noble aspirations. In the ongoing struggle for power, ministries will seize upon any mechanism that can give them an edge against rival bureaucracies.

By identifying seminal technologies for the future and providing seed money, national projects can also help research directors at various companies build a consensus about research directions and R&D priorities. When a critical mass of leading *kaisha* (large Japanese corporations) decides to pursue a basic technological course, other companies are likely to follow. An entire industry can be mobilized. By dividing up the research labor, moreover, national projects can prevent wasteful duplication, without crowding out parallel research efforts in company laboratories. Making project patents available to all companies on a nondiscriminatory basis also ensures that key technologies will be widely diffused, overcoming the barriers to cross-pollination resulting from Japan's relatively low labor mobility.

MITI has organized a variety of national research projects, covering everything from opto-electronics to bioelectronic engineering (see Fig. 2.2). As national projects have proliferated, foreign competitors have expressed concern about the lack of openness and transparency and the asymmetrical advantages conferred on Japanese corporations. This concern implies that national research projects have been and will continue to be highly successful undertakings. Some foreign analysts give them credit for catapulting Japan to the technological forefront.[15] Other see them as vehicles for "unfair" competition, tangible evidence of "Japan, Inc." in action.[16] Many fear that Japan will surge ahead in most fields of high technology, once the multiple projects on the drawing boards or currently under way are completed.

Leaving aside the media attention and expressions of foreign con-

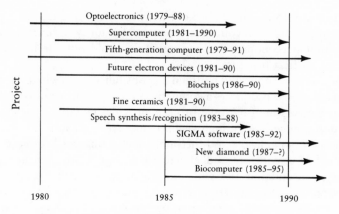

Fig. 2.2. MITI joint R&D projects. Source: DATAQUEST.

cern, how much have national research projects actually accomplished? Probably not as much as foreign Cassandras fear. Such projects, to begin with, are intended to complement, not to replace, corporate R&D. The largest and most important research effort continues to be conducted within the laboratories of individual companies.[17] What national projects provide is a foundation of basic knowledge on which applied commercial development can be conducted. Applied development is still the main driving force behind Japan's commercial competitiveness.

While government subsidies relieve companies of some of the burdens of continuous R&D investment, leading companies are not always enthusiastic about participating in national projects; often they feel they possess more advanced technology than their competitors and believe they have little to gain from joining. Successful organization of national projects is thus dependent on the relatively even distribution of technological capabilities among participating firms. There must also be the expectation that the project will yield substantial collective benefits with regard to foreign front-runners. The paucity of national projects in chemicals, pharmaceuticals, and machine tools—all important high-technology endeavors—can be understood in terms of the absence of one or more of these conditions.

None of the national research projects have yet achieved momentous breakthroughs in state-of-the-art technology. Some have failed to reach even modest objectives. The 3.75 Series Computer Development Project, the first of its kind, was considered a failure from a technical standpoint. Likewise, the Software Development Project (1976–80) failed to fulfill the specific mission for which it had been organized: the development of computer-written applications software. Although an effort was made to cast the results in a positive light by referring to the "library of working aids for programmers" that had been developed, no amount of window dressing could disguise the fact that the project had fallen short of the goals that had been set. Only about 20 percent of the software packages developed during the six-year project have turned out to be commercially useful.[18]

The VLSI project, trumpeted as Japan's greatest triumph, advanced the state of semiconductor technology, particularly Japanese capabilities in process technology (electron beam lithography, silicon crystal growth and processing, device testing, and so forth). But the VLSI project failed to make state-of-the-art breakthroughs, except for its work in the use of liquid crystals.[19] Some people believe that most of its developments would eventually have taken place anyway. All the VLSI project did was to hasten the process, a noteworthy but not earthshaking contribution.

The project's most heralded achievement—collaborative research at four cooperative laboratories involving five participating companies and specialists from MITI's Electrotechnical Laboratories—took several years of administrative massaging by the project's executive director before it could get on with the business of joint research. For the first several years, mutual suspicion and fears about leaking proprietary information impeded the free exchange of information. "What I actually did [the] last four years," says the executive director, "was nothing but to chat with the staff over sake" to create the kind of atmosphere that would be conducive to joint research.[20] The fact that collaboration could be achieved at all should not be denigrated; but whether it was worth the years of trial and error and administrative massaging, or whether the lessons learned can be applied to future projects, is not clear.

The research being done at the Institute for New Generation Computer Technology (ICOT), the central laboratory for Japan's Fifth Generation Computer Project, suggests that meaningful collaboration is repeatable.[21] But since the project is still several years from completion, it is too early to tell whether a major breakthrough will take place. The same holds true of most of the other ambitious projects currently under way. An objective assessment of the technological value of national projects must therefore await the outcome of the current generation of research activities.

Nevertheless, on the basis of the track record so far, some tentative conclusions can be drawn. National research projects have made the biggest contributions in the following areas: (1) identifying the seminal technologies on which R&D cooperation can take place; (2) promoting extensive generation and exchange of information between industry, government, universities, and the financial community; (3) allocating more R&D expenditures and subsidies for private companies, which are especially helpful during cyclical downturns; (4) helping companies commit themselves to the long-term development of vital precommercial technologies; (5) transferring know-how from government to corporate laboratories; (6) encouraging and facilitating close contact among scientists and engineers; (7) diffusing precommercial technology throughout an economy where career-long employment limits the speed and scope of diffusion; and (8) equalizing technological capabilities among leading firms and intensifying the race to develop new products and process technologies. Of these contributions, the last four have been of particular importance.

The list suggests that, quite apart from the question of breakthroughs, national research projects appear to serve a variety of instrumental functions. Indeed, the value of national projects may reside as much in their secondary effects—corporate commitments, technological diffusion, and stepped-up market competition—as in their primary goals—advances in state-of-the-art technology. Assessments of their long-term effectiveness must take both primary and secondary effects into account.

Although national research projects have hastened the pace of latecomer catch-up, it is far from clear that they will continue to impart as much impetus in the future. When Japan was a latecomer, the research directions to take were fairly clear. All that was required was to observe the successes and false starts of the front-runners. Now that Japan has come to the frontiers of technology, the natural paths to take are no longer clear. If MITI sets industry off in pursuit of the wrong set of technological goals, or charts the wrong pathways to discovery, the miscalculations might be exceedingly costly—not only in terms of the investment of financial and human resources but also in terms of lost time and forgone commercial opportunities. Setting technological targets has thus come to involve greater uncertainties and risks.

The risks of moving in false directions, however, are more effectively minimized in Japan's system of industrial policy-making than in most other countries, owing to the scope and depth of government consultations with the private sector. As pointed out in the last chapter, MITI engages in painstaking discussions with scientists and engineers, research scholars, industry leaders, and financial analysts—the people in the know—to find out where technology is headed and where the most promising commercial opportunities lie. The information it collects and processes is about as thorough as could be obtained. National research projects thus emerge from an ongoing process of national consensus building based on extensive give-and-take between government and the private sector.

NTT-Related Research

The visibility of national projects should not cause one to lose sight of important research simultaneously going on elsewhere in Japan, at such places as Nippon Telegraph and Telephone's (NTT) four large laboratories, the Electrotechnical Laboratories (ETL), the Science and Technology Agency, and other research institutes. From the standpoint of research in microelectronics, these government and public corporation laboratories—working in close contact with researchers from the private sector—have probably done more to raise the level of Japan's technological capabilities than any national project completed as of 1988. In the early 1950's, the Electrical Laboratory (Denki Shikenjo), ETL's predecessor, and the Electrical Communications Research Laboratory (Denki Tsūshin Kenkyūjo), forerunner of the NTT labs, laid much of the groundwork for Japan's introduction of transistor technology. The first transistor ever made in Japan, the point contact transistor, was developed at the communications lab in October 1951.[22]

NTT operates four major laboratories, employing over 3,000 scientists, engineers, and researchers, and is planning to establish another research lab in the Kansai region. NTT had a budget of ¥94 billion in 1983, or about $400 million.[23] Of that amount, NTT spent about ¥35 billion ($146 million) for four major projects: information processing (¥12 billion), digital switching (¥8.1 billion), large-scale integrated circuits (¥8.7 billion), and satellites (¥6.4 billion). Owing to its superb research facilities, large R&D budget, and prestige, NTT has been able to attract top-notch college graduates and conduct research of the highest quality in such fields as integrated circuits, electronic switching, power transmission, and data processing. Its laboratories have played a central role in advancing Japan's technology; NTT holds over 8,000 patents.

During the early decades of the postwar period, the quality of NTT research in microelectronics was a cut above that conducted in corporate laboratories. NTT worked closely with family firms across a broad spectrum of technical tasks, contracting out research, exchanging information, and joining forces with them on certain technical problems. Over the years, a number of NTT research specialists, like higher civil servants, "descending from heaven" (*amakudari*) into high-level positions in the private sector, have left NTT to assume positions of high responsibility in the R&D divisions of leading NTT family firms. The close relationship has thus functioned as a conduit for the transfer of technology from NTT to the private sector and vice versa, enriching the state of knowledge on both sides.

Japan's capacity to overwhelm U.S. manufacturers of mass memory chips has been due, in no small measure, to joint research conducted by NTT family firms with NTT and the diffusion of NTT technology. NTT laid important groundwork in the development of the 64K random access memory (RAM), the 256K DRAM, and the 1M read only memory (ROM), covering device designs and production technology. Japanese manufacturers could not have come from so far behind in such a short time without NTT's diffusion of technology.

To get a notion of what NTT has meant to the Japanese microelectronics industry, imagine the impetus U.S. companies would have received if AT&T had purchased all its supplies from the merchant market and if U.S. firms had been given the same opportunity to work closely with Bell Laboratories as Japanese firms had with NTT labs. Because AT&T's needs were met internally by Western Electric, its captive supplier, however, the direct benefits for the U.S. electronics industry turned out to be small. Nor did Bell Labs work in as close physical proximity with leading private companies as NTT did, even though top researchers left Bell Labs to join private-sector companies.

To be sure, Bell Labs had to make its patents freely available at reasonable fees as part of the 1956 antitrust consent decree. The effect of this transfusion of technology on the growth of the U.S. telecommunications, semiconductor, and computer industries can scarcely be overstated. But Bell Labs' compulsory licensing not only invigorated the U.S. information industries but also facilitated the growth of Japan's fledgling information industry, because Bell patents were also available for international transfer, and Japan took full advantage of the available know-how.[24]

The consequences of the divergence of U.S. and Japanese telecommunications structure and antitrust policies—each put in place independently on the basis of domestic considerations, with little or no thought being given to the international repercussions—were entirely fortuitous but far-reaching. AT&T's structure and the international diffusion of Bell technology kept the barriers to new entry low not only for U.S. companies but also for Japanese and other foreign late-comers. NTT's structure, particularly the absence of a counterpart to Western Electric and the intimate give-and-take relationship between NTT labs and private Japanese companies, provided Japan's private producers with the opportunity to speed up the process of industrial catch-up. Japan benefited from the United States' open structure and strict antitrust enforcement. Had the situation been reversed, with Japan and NTT the pioneers and the United States and AT&T the latecomers, the U.S. information industries would have had a much harder time overtaking the Japanese front-runners.

The flow of benefits has not been entirely one-way. NTT family firms have shared proprietary technology with NTT. One Japanese company, which shall be called Daimaru, started work on fiber optics and gallium arsenide during the late 1960's, when most companies were hesitant to invest in these unproven technologies. The early start and sustained attention put Daimaru well ahead of everyone else. When fiber optic and gallium arsenide products became commercially feasible, Daimaru willingly shared its know-how with NTT on a con-fidential disclosure basis. Then, with its circle of family firms, NTT succeeded in developing the vapor axial deposition (VAD) fiber optic production method, considered the most advanced in the world.[25] To standardize fiber optic cables, NTT licensed all Japanese cable makers to use VAD, thus diffusing the technology.

Daimaru was willing to disclose what it had learned through fifteen years of painstaking research because its relationship with NTT was of overriding importance. The relationship can be accurately described as give-and-take reciprocity based on mutual trust and long-term com-mitment, an illustration of the infusion of organization linkages into market transactions.[26] Like other family firms, Daimaru knew NTT would safeguard the confidential information, even though it realized that some portion would eventually be recycled to competitor firms. To begrudge information out of short-run fears of leakage would have violated the spirit of trust and obligated reciprocity on which NTT–family firm relationships are based.

Here is an illustration of what is organizationally fairly common in Japan, a long-term strategy of variable-sum cooperation.[27] Obligatory reciprocity is common among firms that are bound together by ties of structural interdependence, such as parent corporations and subcontractors and keiretsu companies.[28] What is unusual is that confidentiality and trust also extend to relations between the private and public sectors.

In addition to its bilateral interactions with individual firms, NTT also organizes its own version of multifirm research projects. In the mid-1970's it sponsored its own VLSI project in parallel with the better-known MITI undertaking by the same name. NTT's VLSI project was oriented toward applications in telecommunications equipment while MITI's concentrated on computer applications. NTT has coordinated R&D efforts in key areas of telecommunications technology such as digital switching systems, an area requiring large R&D outlays, high risks, common standards, and the development of compatible equipment.[29]

Until NTT's privatization, the old system of NTT-related R&D may have given more momentum to the development of microelectronics technology than MITI's national research projects. NTT family firms became very competitive under NTT's auspices. For the microelectronics industry as a whole, the old system reduced risks, lowered costs, and set very exacting standards that helped to upgrade overall quality in Japan's commercial marketplace.

The combination of NTT and national research projects, together with research conducted at other government laboratories (like those at Tsukuba), has helped to compensate for past shortcomings in Japan's university-based system. What is striking about both NTT and the national projects is the synergism—not simply the spillover benefits—produced by the interplay between joint R&D efforts in the public and private sectors. Through such interaction, public and private interests converge and the collective good is advanced.[30]

Government Financing

As pointed out earlier, one of the key functional requisites of high technology is that companies sustain high levels of capital investment, including the escalating costs and risks of R&D. As part of its strategy of "technology push," the Japanese government has tried to lighten the burden of high-tech companies by making low-cost capital available. During the 1950's and 1960's, when investment capital was in short supply, the government played a pivotal role in allocating capital to the high-priority industries. But the importance of this role has decreased over time as investment capital has become plentiful.

The proportion of funds available to industry through the Fiscal Investment and Loan Program (FILP), the main source of public funding (from postal savings and insurance funds), has declined steadily since the early 1950's. FILP funds accounted for nearly 30 percent of the total capital available to industry in the early 1950's, but its share in the 1980's has fallen below 10 percent, with only a tiny portion of that going to high technology. According to one estimate, the government's share of plant and equipment investments in the electronics equipment industry came to only 2.5 percent during the early 1960's and a mere 0.8 percent in the late 1970's.[31] Compared to the amount the French government is pouring into its electronics industry, this is modest indeed.

Since Japan's cup of investment capital is now running over, there is much less need to channel public funds to priority sectors. Even the practice of indicative lending—the funneling of small amounts of public funds to targeted industries, prompting private institutions to follow suit—has diminished as an instrument of capital allocation. Leading corporations in targeted sectors are fully capable of meeting their own needs through a combination of retained earnings, bank loans, and stock and bond issuances.

Research Subsidies

In what forms, and to what extent, has the Japanese government subsidized research in the information industries? The perception outside Japan is that the subsidies have been huge—far out of line with what governments in other countries provide. Some believe that a case could be made that Japan has violated the Subsidies Code of the Tokyo Round agreement.[32] Is this the case? Does the Japanese government give larger subsidies than the governments of other countries?

In providing funds for research, the government can draw upon several sources: special grants allocated by the Ministry of Finance (MOF) for national research projects, internal allocations from MITI's own budget (including "hidden" funds available from energy-related taxes and bicycle racing proceeds), and support from other government ministries (Science and Technology, Education, Posts and Telecommunications, and the Japan Defense Agency). Let us calculate the total by adding up the known categories of expenditures, beginning with the budgets for national research projects.

Government subsidies for national projects can be grouped into two periods: the era of frenetic catch-up (1970–79) and the period of

TABLE 2.1

Government-Supported Research Projects in Japan, 1966–1980

Period	Project	Amount (million $)
1966–71	High-performance computer R&D	$42
1972–76	3.75 Series Computer development	228
1971–80	Pattern information processing system	115
1976–80	VLSI development	150
1976–80	Software development	30
TOTAL		$565

SOURCE: Personal interviews.

state-of-the-art development (1980–90). Government outlays from 1966 to 1980 amounted to around $565 million, or roughly $43 million per year; this represented less than 10 percent of the total R&D expenditure for the information industry, the rest of which was shouldered by private enterprise. The main research projects of this period are listed in Table 2.1. Public funds were small, whether they are measured in the aggregate, averaged out on an annual basis, or calculated as a proportion of total R&D in the information industry.

In the United States, the government has funneled far larger aggregate, annual, and percentage sums into R&D. In 1957, for example, the Air Force, Army, and National Aeronautics and Space Administration (NASA) provided $518 million for electronics and communications equipment, or roughly 70 percent of the total spent on this industry. Even as late as 1968, the U.S. government accounted for about $1.5 billion, or roughly 60 percent of total R&D.[33] It should be pointed out, of course, that the U.S.-Japan comparison is a bit misleading, given the fact that the United States served as the technological pioneer and Japan as the follower. For firstcomers, the costs of R&D are usually far higher than for latecomers, if only because the level of initial uncertainty is so much greater; failures and false starts are almost unavoidable. Latecomers, knowing what has worked and what has not, have a decided advantage in that they deal with less uncertainty. Often key patents can be obtained at a fraction of the cost that would have been incurred. Nevertheless, the comparison brings to light the relatively modest sum of subsidies provided by the Japanese government from the mid-1960's to 1980.

The small sum shrinks ever further when one notes that most of the early subsidies were conditional loans, or *hojokin,* repayment of which was contingent on the success of the project. Although conditional loans would seem to create perverse incentives to "cheat" on repayments—what Oliver Williamson calls "opportunism" or "self-seeking with guile"[34]—a surprisingly high percentage of loans have been repaid. Of the hojokin grants made by MITI's Agency for Industrial Science and Technology (AIST) over the five-year period 1974–78, nearly half, or 43.6 percent, had been repaid by 1982.[35] This record of repayment thus reduced the net R&D subsidy to an unspecified amount well below the aggregate figure cited for national research projects. (The reduction cannot be specified because the data on repayment are not broken down by research project, making it impossible to ascertain which of the information industries projects had been repaid.) The amount can be estimated roughly in terms of the unpaid principal plus the uncharged commercial interest rates.

MITI chooses which projects to organize and subsidize very carefully, in close consultation with industry. To qualify for government assistance, a project must meet four criteria: (1) the proposed project must be of seminal importance to Japan's technological development and future economic well-being; (2) the research must be of a precommercial nature so that participating companies do not gain a decisive commercial headstart over excluded firms; (3) government assistance must be indispensable for the project to get under way and be completed; (4) the time frame for the project's completion must be realistic. Government financing provides the critical missing ingredient for launching projects of high capital costs and risks, relatively long gestation, fundamental technological importance, and broad commercial applicability.

MITI prefers to put up only a portion of the total capital necessary to finance a national project.[36] Usually the sum is less than half the total cost. The rationale is that the subsidy should represent seed money, a sum that makes the project feasible but requires industry to put up its own capital. If MITI bore the full costs and risks, it might dampen industry's incentive to be efficient. MITI might then be strapped with the same problems that plague DOD research in the United States: padded expenses, unanticipated delays, huge cost overruns, and failure to meet strict technical specifications and development targets. Other things being equal, therefore, MITI's clear preference is to insist on cost sharing through the extension of hojokin, low-interest loans repayable from the profits made on commercial products.

TABLE 2.2

Government-Supported Research Projects in Japan, 1980's

Period	Project	Amount (billion ¥)
1979–83	Software for VLSI hardware	22.5
1976–82	Software production technology	6.6
1979–86	Optoelectronics applied system	18.0
1981–89	Fourth Generation high-speed computer	51.5
1981–91	Fifth Generation computer	10.5*
1981–90	Next-generation industries technology	25.0
1977–84	Flexible manufacturing	13.0
Continuous	Important technology	2.1
1983–90	Critical work robot	17.5
TOTAL		¥ 166.7

SOURCE: U.S. Embassy, Tokyo, unclassified telegram, May 1982, cited in Arthur D. Little, Inc. (Japan), "Summary of Major Projects in Japan for R&D of Information Processing Technology," unpublished study, 1983.
*Amount for 1981–84 only.

Having closed the gap with U.S. front-runners in most areas of high technology, the Japanese government turned its attention to the organization of more ambitious national research projects aimed at achieving state-of-the-art breakthroughs. The 1980's have witnessed the organization of nine major projects related to the information industries, all designed to propel Japan beyond the frontiers of knowledge (see Table 2.2).

Unlike most national projects during the catch-up phase, many of the state-of-the-art projects listed in Table 2.2, with the notable exception of the Fourth Generation computer project, involve the government's assumption of all expenses. Such financing, referred to as *itakuhi*, is akin to contract research in the United States. Itakuhi has come to be used more than hojokin because state-of-the-art projects are by definition substantially more costly, riskier, more uncertain, and of longer gestation. Private companies are more hesitant to commit their own resources.

For several of these ambitious projects, MITI has had to take the initiative, instead of simply responding to industry demands. Had it not done so, the projects might never have been launched. Private firms would not have been willing to take the initial step of bearing the full costs of precommercial research. An employee of one of the companies that agreed to participate in the Fifth Generation project explained the financial arrangements from the private sector's point of view:

At first, MITI wanted to support this project at only 50 percent for the first three years, with private firms supplying the other 50 percent of the funding, but we in the companies said no. We can't afford to support such a high-risk project, even at 50 percent, plus contribute researchers' time. When they saw we meant it, they agreed to support it 100 percent, at least for the first three years. After that, we'll see.[37]

The shift from hojokin to itakuhi has increased the actual amount of subsidies significantly. Over 60 percent of R&D subsidies from 1976 to 1982 were hojokin, subject to eventual repayment. But for the nine projects listed in Table 2.2, itakuhi support has come to surpass hojokin by a wide margin.

The government's willingness to underwrite the full cost of a number of the frontier R&D projects, however, should not be misinterpreted. It does not mean that all private companies are eager to jump on the bandwagon and reap the benefits of the government's free ride. Consider the case of the Fifth Generation project; several firms had to be coaxed into participating, and several groused openly about it. On their visits to Japan, Edward A. Feigenbaum and Pamela McCorduck observed that "resentment and hostility are hardly strong enough to describe the attitudes of another firm's managers toward the Fifth Generation. They told us frankly that they had not wanted to participate and only under duress (whose nature we couldn't ascertain) did they finally contribute their researchers to ICOT."[38]

Quite apart from having to divert research personnel, Japanese corporations often have the same reservations about itakuhi as U.S. companies do about federal contract research: reams of paperwork, minute and irritating regulations, rigid accounting procedures, constant government monitoring, strict technical specifications, no guaranteed markets for commercializable products, and so forth. Sometimes it does not seem worth the hassle.

On top of the bureaucratic red tape, contracting companies in Japan are usually not allowed to retain proprietary rights over the research results. Patents automatically revert to the sponsoring government agency, which makes them available on a nondiscriminatory basis to all nonparticipating companies for the cost of a patent license fee. Despite full financial coverage, therefore, itakuhi is not as attractive to private companies as the term *free ride* might suggest. The administrative costs alone are hardly trivial.

The aggregate sum for all Japanese national research projects—

¥166 billion spread over more than ten years—sounds enormous, and the figure does not include costs for the second half of the Fifth Generation computer project. Nor does it include internal MITI funds requiring no Ministry of Finance authorization, such as revenues from bicycle racing and special energy tax revenues. In 1982, for example, MITI was able to allocate around ¥2 billion (about $8.5 million) from its regular budget for high-tech R&D support. In addition, discretionary funds are available every year from revenues gained from regulated gambling at bicycle races. The amount varies from year to year, depending on the amounts gambled, but in 1982 it came to around ¥27 billion (about $112 million), or slightly more than 5 percent of MITI's budget. Most of that money is used to support miscellaneous activities such as trade fairs, public relations activities, and trade associations, but some can be expended to underwrite research in advanced technology.

From MITI's standpoint, the advantage of drawing on bicycle racing revenues, instead of special grants from the Ministry of Finance, is that administrative entanglements can be averted. MITI does not have to go through the time-consuming process of submitting formal budget proposals; there is flexibility in the choice of technological focus, and companies are not saddled with bothersome reporting requirements. MITI has a free hand. In 1982, it allocated ¥800 million (about $3.3 million) for research in computer software and data processing, an area of pressing need in which Japanese companies were lagging behind U.S. pacesetters.

The real value of discretionary funds is not so much the incremental amounts that are available, but the flexibility they bestow. MITI drew on these funds, for example, when the Ministry of Finance balked at the idea of underwriting the risky Fifth Generation computer project. By putting up its own funds, MITI was able to get the project launched. Once its feasibility and long-term value had been demonstrated and both momentum and excitement had been generated, the project received substantial funding from the Ministry of Finance. Without the discretionary funds at MITI's disposal, the Fifth Generation computer project, Japan's best-known and perhaps most important project to date, might never have moved beyond the planning stage.

If we include a portion of NTT's R&D expenditures, the grand total for research subsidies over a ten-year period comes to ¥516 billion (roughly $2.3 billion at an exchange rate of 230 yen to the dollar, the official rate in September 1985). Although the figure seems enormous, it averages out to only ¥51.6 billion per year over a ten-year period ($230 million per year). For any given year, that figure represents only a fraction of national R&D expenditures. Compared with Bell Laboratories' 1983 budget of $2 billion[39] or IBM's yearly budget of over $1.6 billion,[40] Japan's $230 million seems modest; keep in mind also that the latter figure includes the principal of conditional loans that have to be repaid and thus overestimates the actual subsidy. This brings us back to the point made earlier: the purpose of government subsidies is to compensate for market imperfections by serving as catalysts, not substitutes, for private-sector investment in basic and precommercial research.

U.S. and European Research Subsidies

What about the U.S. government's support for R&D in the information industries? How does it compare with Japan's? Let us look at only a few of the federal government's most highly publicized projects, most of them sponsored by the Department of Defense (DOD). The very high speed integrated circuit (VHSIC) project, launched in 1980 and scheduled for completion by 1989, calls for DOD to put up an estimated $500 million in contract research.[41] In partial response to Japan's Fifth Generation project, DOD has also launched its own Strategic Computer Project, designed to advance the frontiers of artificial intelligence, software, and computer architecture, with a projected budget of around $600 million for the first five years, and perhaps another $900 million for the final five years. Two other DOD projects of note are a $100 million program for the development of gallium arsenide circuits and a seven-year, $250 million project called Software Initiative. The allocations for just these four projects and U.S. government financing come to $2.35 billion, roughly the same as the aggregate sum in Japan, including the Japan Development Bank, MITI, and NTT. Since virtually all the money is in the form of contract research (none of it in repayable loans), the actual amount of the subsidies in the United States for the four projects may exceed Japan's total over a ten-year period. If subsidies for all other projects were included (for which full statistics are not available), the U.S. level would exceed Japan's by several times.

The U.S. government accounted for over 18 percent of total R&D expenditures in the category of office machinery (SIC code 357) and 40 percent in communications equipment (SIC code 361–64, 369) in

1980.[42] These figures exceed those in Japan by a big margin. But it should be pointed out that most of the U.S. government money goes to support military-related R&D projects, with only limited spillover benefits for commercial applications. From a purely commercial point of view, therefore, the U.S. government total is not nearly as cost-effective as Japan's; to the extent that it is not, the dollar amounts of federal subsidies should be discounted. Nevertheless, U.S. and European criticisms of Japanese subsidies—even allowing for low commercial spillovers—sound hollow or hypocritical.[43]

Certain countries in the European Community also provide larger research subsidies than Japan. Take, for example, semiconductor research: state financing in the United Kingdom topped Japan's.[44] The government in France has also subsidized its electronics industry more heavily. Under President François Mitterrand, the French state embarked on an ambitious crash program to upgrade France's capabilities in electronics and other high-tech industries. It set aside $1 billion for investment in France's electronics industry over a five-year period, 1982–86, with $600 million earmarked for R&D and $400 million for expanding manufacturing facilities.[45] The $1 billion was only a fraction of the massive $17 billion package in public and private funds that Mitterrand has committed to the development of electronics. Nor is that all. The troubling inability of the French computer industry to compete effectively against U.S. and Japanese companies has prompted the French government to take the ultimate step: nationalization of major parts of the computer industry and its organization into four autonomous subsidiaries under the control of one holding company, Compagnie des Machines Bull.[46]

On top of subsidies supplied by individual countries, the European Commission (EC) has launched a major five-year research project called Esprit, which is aimed at raising Europe's level of technology in microelectronics, software, artificial intelligence, office automation, and computer-aided design and manufacturing (CAD/CAM).[47] The EC is also seeking to streamline research across the European continent in order to achieve the benefits of scale economies and to curtail the waste that results from excessive duplication of R&D. By one estimate, the total amount of money, public and private, invested in European R&D in 1982 was more than double that invested by Japan.[48] Should projected spending levels stay on track until 1990, the gap will widen. Such spending levels are high relative to the overall size of Europe's electronics industry. Europe's share of the world market for integrated circuits in 1982 came to a mere 7 percent, and

European-built computers accounted for only 10 percent of world production. State subsidies in Europe as a percentage of total electronics output, therefore, surpass those in Japan by a wide margin.

Whether measured domestically, in terms of total investment, or internationally, in comparison with U.S. and European outlays, Japanese government subsidies cannot be considered abnormal. It is, however, difficult to standardize measures for international comparisons. How can Japan, the United States, and Europe be compared when there is not even agreement on what constitutes a subsidy? Is it best to add up loans, outright grants, and contract research? How does one adjust for differences between military and civilian research or for interest rate differentials? How can adjustments be made for fluctuations in official exchange rates?

Even the measure referred to as the "marginal subsidy equivalent"—the value obtained when a company produces an additional unit of output as a function of a given subsidy input—is beset with problems.[49] What we are left with, then, is rough, aggregate data that provide only crude evidence of government R&D assistance. The weight of the evidence indicates that the Japanese state is hardly alone in thinking that the market mechanism must be supplemented with subsidies if society is to reap the full fruits of innovation and growth in high-technology industries.

If anything, the level of Japanese subsidies is located toward the lower end of the bell-shaped curve of world distribution, with the United States, France, and the United Kingdom nearer the higher end. It is too early to tell whether the massive government and EC subsidies will propel Europe to the technological forefront. If past patterns hold true, the likelihood is that simply pumping more subsidies into the R&D effort will not remedy whatever it is that ails Europe. Although technology push by the government appears to be a necessary condition of leadership in high technology, it is clearly not a sufficient condition. Commercial competitiveness hinges on far more than the amount of government largess made available. Indeed, the cost-effectiveness of subsidies can no doubt reach a point of diminishing returns. Used to excess, they may even dull incentives to compete.

The effectiveness of government subsidies depends less on their size than on the intangible factor of the private sector's capacity to convert public assistance into technological progress and commercial competitiveness. Japanese industry has demonstrated time and again that it possesses the capacity to derive the most from government assistance.

In many contexts and in a variety of industries, the Japanese have demonstrated an unsurpassed talent for collective learning, an ability to organize themselves effectively for such diverse tasks as research, production, and marketing.[50] It is this collective learning capacity, not the size or number of subsidies, that accounts for Japan's success at making the most of government seed money.

All governments in advanced industrial countries subsidize their high-tech industries in one form or another; only the nature and scope of subsidization vary. The need for government support can be understood in the context of the characteristics and functional requisites of high technology, discussed earlier, and the very high stakes riding on the outcome of international competition. For lagging countries, reliance on research subsidies might be preferable to import protection. Recall that, in Japan's case, R&D subsidies rose substantially when foreign pressures forced a lowering of tariff and investment barriers. When faced with a hard choice between subsidizing and protecting domestic producers, many governments will opt for subsidization as the less costly, more palatable course of action, particularly in terms of the norms of free trade. The problem is that some governments may feel compelled to combine subsidies with trade protection.

Other Policy Instruments

Tax Policies

For most governments, taxation is perhaps the most readily available instrument of industrial policy. Tax provisions that discriminate between sectors, for example, offer an immediate means of providing special encouragement to priority industries. From a political standpoint, the virtue of taxation is that it requires no direct or visible drain on government budgets. This means that it can be shielded from the glare of public scrutiny and more effectively insulated from the pressures of political accountability. It is a ubiquitous tool of industrial policy, politically easier to use than most line-item, zero-sum budget allocations, for which special-interest lobbying can be ferocious.

The danger is that tax policies, indiscriminately used, can become the handmaiden of political expediency, with short-term concessions to vested interest groups exacting heavy costs in terms of long-run economic efficiency. The problem is particularly acute when the interest groups with the most political clout also happen to be among the most inefficient economically, as is often the case in advanced industrial countries. Old-line industries, which employ large, unionized labor forces, tend to maintain their political influence long past their economic prime. There is a lengthy time lag between the retention of political power and the loss of economic viability, an "iron law of rigidity,"[51] so to speak, restricting the ability of advanced industrial economies to adapt their industrial structures to changes in the international division of labor.

The power of old-line industries helps to explain why, in the United States and most European states, intersectoral tax burdens are heavily skewed in favor of economically inefficient sectors. Reflecting the organized strength of old-line interest groups, uneven tax burdens have the effect of channeling resources to the declining sectors of the economy. This distorts the government's revenue base and shifts the weight of taxes onto the shoulders of the most efficient and promising sectors, a perverse reward for efficiency and growth.

Even when the U.S. Congress is moved to pass generous tax provisions for high-tech-related R&D, such as capital gains, investment credits, and accelerated depreciation allowances, the provisions can lead to unforeseen and sometimes undesirable consequences. To begin with, Congress is apt to extend such provisions to other sectors in order to avoid the appearance of playing favorites. Even if it does not, certain non-targeted interest groups, like real estate, may wind up reaping the biggest windfall benefits; this may represent a substantial cost to the U.S. economy in terms of lost tax revenues, rising real estate prices, and allocative distortions caused by misinvestments in the less productive sectors.

There is, thus, the danger that the legislative branch will treat inefficient industries overgenerously on two fronts: (1) through the uneven distribution of corporate tax burdens, and (2) through the extension of special tax incentives, which benefit the less productive sectors disproportionately. A potentially effective tool of industrial policy, therefore, can turn out to be a blunt and harmful instrument for high-tech industries.

The relative weakness of the legislative branch in Japan and MOF's policy-making authority over tax measures have produced a more coherent, less distorted set of tax policies. The declining, inefficient industries have failed to secure blatantly inequitable tax concessions. The corporate tax burden is more evenly distributed among industrial sectors in Japan than in either the United States or the United Kingdom.[52] The even distribution means that the targeted sectors receive less preferential treatment than is commonly presumed. But, since old-line industries also receive less favorable treatment, the high-tech sectors may be better off overall, by virtue of the smaller distortions caused by preferential tax policies.

According to a study by Gary Saxonhouse, the effective rate of capital taxation in Japan in 1973 ranged from 34.7 percent (nonferrous metals) to 49 percent (electrical machinery); in the United States, the variation was much wider, ranging from 19.7 percent (petroleum) to 131.2 (electrical machinery).[53] Yet the average corporate tax rate, including state and local levies, was roughly the same in both countries: 53.2 percent in Japan, compared to 51.2 percent in the United States. Thus, targeted taxation is, in some senses, practiced less widely in Japan than in the United States; the declining and inefficient sectors tend to be the biggest beneficiaries of preferential tax treatment in the United States. The differences can be attributed largely to regime characteristics, particularly the greater power exercised by the legislative branch in the United States.

In Japan, the two key bureaucracies, MOF and MITI, enjoy greater insulation from the pulling and hauling of interest-group politics. Every year, MOF and MITI negotiate an aggregate ceiling for special tax measures, freeing MITI to grant special tax exemptions in whatever amounts it deems appropriate for industries of its choosing (so long as it stays within the limits of the agreed-on aggregate ceiling). Various divisions and bureaus within MITI vie with one another to win special tax provisions for industries under their jurisdiction. The contending claims are aggregated, hard trade-offs are made, and a unified package of special exemptions is worked out within the ministry. The Business Behavior Division (Kigyō kōdōka) of the Industrial Policy Bureau (Sangyō Seisakukyoku) decides which industries deserve to receive tax exemptions, and in what amounts, after consulting with all the sector-specific divisions and bureaus and huddling with the Accounting Division (Kaikeika) and the General Coordinating Division (Sōmuka) of the Minister's Secretariat (Daijin Kanbō).

In aggregating competing interests within MITI, special tax provision are one of four related policy instruments; the others are subsidies, legislation, and administrative guidance. The General Coordinating Division tries to blend all four ingredients, in appropriate measure, into a recipe that meets the needs and peculiar circumstances of each industry. Thus, declining industries may get a stronger dose of administrative guidance and subsidies, while for high-tech industries, tax incentives and research support may receive more weight. Whatever the mix, MITI sees the four elements as related and to some extent interchangeable parts of a whole.

The virtue of Japan's system is that it (1) keeps a cap on the estimated losses due to special tax incentives; (2) forces all industries to compete with each other for special tax treatment on the basis of what is in the best interest of the industrial economy as whole; (3) gives MITI the leeway and authority to determine the optimal uses of special tax incentives; (4) holds parochial politicking in check; and (5) balances the use of tax measures with other tools of industrial policy. Tax targeting is thus subsumed within the broad framework of other policy measures, not utilized in isolation as an elastic solution to short-term political lobbying. That Japanese industrial tax policies come out bearing fewer signs of obvious inconsistencies than U.S. policies is therefore hardly surprising.

With respect to specific tax incentives, such as the one for R&D investments—a matter of obvious importance for high-tech industries—the stereotype of unfair Japanese tax provisions is again belied by comparisons with the United States. Japan grants tax credits of 20 percent for all R&D expenditures that exceed the highest annual rate in a corporation's past, up to a ceiling of 10 percent of the corporation's taxes. The United States, by contrast, allows tax credits of 25 percent for all R&D expenditures exceeding the *average* over the preceding three years; there is no ceiling on the amount that is deductible and the tax credit can be carried over a fifteen-year period. Japanese corporations saved ¥27 billion (roughly $122 million) in 1981, thanks to this R&D tax credit; no comparable data were available for the United States, but given the higher tax credits and the larger amount of R&D investment, the figure surely exceeded that.

In the United States, high-tech industries also receive investment tax credits of up to 6 percent of the value of new equipment on depreciation schedules as short as three years, and up to 10 percent for equipment with a longer life. Japan offers nothing comparable. Such provisions indicate that Japanese high-tech companies receive less preferential treatment than is widely assumed. The situation fits with Saxonhouse's findings about the comparatively low variation in tax burdens across sectors.

This is not to say that high-tech companies in Japan enjoy no tax advantages whatsoever over their U.S. competitors. The most significant advantage is that the effective tax rate on Japanese corporations, taking into account national price factors, is lower than comparable rates for U.S. companies, thanks to lower inflation rates.[54] The high rate of inflation in the U.S. from the mid-1970's to the early 1980's has had the effect of raising real corporate taxes. The negative interaction between inflation and taxes has also inhibited capital invest-ment in the United States. However, the difference between Japan and the United States stems from macroeconomic, not industrial, policies and is systemic in nature, not the result of industrial targeting. In national tax policies, as in R&D subsidies and national research projects, industrial policy per se does not appear to confer outra-geously unfair advantages on Japanese high-tech companies. To the extent that they exist, the discrepancies in Japan's favor can be better explained by reference to regime characteristics than to aberrant pol-icy practices.

Antitrust

Two other areas in which Japan differs from the United States are the enforcement of antitrust laws and the use of administrative guid-ance as tools of industrial policy. Although antitrust is not usually considered a part of industrial policy, the strictness or laxity with which it is enforced has a bearing on the government's flexibility in making use of certain policy tools.

National research projects, for example, would pose greater prob-lems of organization in the United States because of the stricter inter-pretation of antitrust and greater concerns about possible anticompet-itive effects. Japan's Fair Trade Commission (FTC) seems to feel that national projects pose little threat to market competition as long as most of the major firms participate. If participation were limited to a single national champion (as in France and some of the smaller Eu-ropean states), the excluded companies might be placed at an insur-mountable disadvantage; this could bring about unacceptable levels of market concentration and create formidable barriers to new entry.

Inviting most of the large corporations to participate, on the other hand, might give rise to a different set of problems. It might, for example, strengthen the hands of already dominant firms, accentuate oligopolistic patterns of behavior, and relegate other firms to second-class status. Why do the second-tier companies in Japan's information industries—like Sanyo, Sharp, and Sony—accept their exclusion from many national research projects? Why do they not voice objections or seek legal injunctions to prevent oligopolistic practices from taking place, as excluded companies in the United States would be apt to do? Are there not legal grounds for charges of collusion in restraint of trade?

If there are negative antitrust implications, the Japanese government does not appear to be too concerned. Levels of market concentration in high-tech sectors—though higher in some cases (such as semiconductors) than in the United States—remain sufficiently low that the FTC does not have to worry about the dangers of oligopoly. Since demand is still ascending steeply—with new products continually hitting the market and lots of leeway for both new venture start-ups and horizontal entry from companies in related industries—fears of excessive market concentration appear to be unfounded.

In its less stringent enforcement of antitrust, Japan resembles France, Sweden, and other European states more than it does the United States. The difference is that the domestic markets of the individual European states tend to be significantly smaller and the number of domestic producers in any given sector of high technology is usually small. Countries with tiny domestic markets and leeway for only one or two domestic producers feel they cannot afford the luxury of adhering to the strict standards of antitrust enforcement in the United States. It is all they can do to maintain the existence of one or two companies in niche markets. Neither the United States nor Japan has a problem with an undersized domestic market; yet Japan is not as fervent in the enforcement of antitrust as the United States, in part because its government is less preoccupied with the concept of individual consumer sovereignty and more concerned about the collective good.

The basic, precommercial nature of research also makes national projects palatable from an antitrust point of view. Similarly, the availability of patents on a nondiscriminatory basis to all firms neutralizes the most serious objections that excluded companies might raise. The principal advantages gained by participation in national projects are therefore confined to two areas: government funding, which can be used to supplement the firm's own R&D investments; and the benefits of hands-on laboratory experience and close contact with researchers from government labs and other firms, which far exceed the know-how obtainable through access to patent licenses. Although these advantages are certainly significant, they are not so decisive that firms gain an insurmountable edge by participating. Since the selection of participating firms is done on the basis of merit, the second-tier companies express few complaints about being excluded from the projects (though companies like Sanyo and Sharp have been included in some of the recent national projects).

It should be pointed out, furthermore, that U.S. antitrust policies do not rule out all forms of cooperative research. Every year a few dozen requests, usually involving small firms, to engage in joint research are approved. So long as cooperation does not seriously inhibit competition, joint research can be accommodated under the provisions of U.S. antimonopoly law. In DOD projects, collaboration is sometimes encouraged. The VHSIC project, for example, organized companies into several distinct contract teams in an attempt to bring about an efficient division of labor and encourage research cooperation. Hughes, RCA, and Rockwell standardized their computer-aided design system and exchanged information on mask designs, design rules, and patents as they worked to develop complementary metal oxide semiconductor/ silicon on sapphire (CMOS/SOS) technology.[55] Perkin-Elmer and Hughes Research Labs also joined forces in the development of electron beam lithography, exchanging very sensitive company information in order to achieve better results. Collaboration was also necessary to ensure compatibility for key technological systems such as the signal processor chip set. Thus, interfirm collaboration and cooperation do take place in the United States under the aegis of government projects, though far less extensively than in Japan.

What about interfirm cooperation in the nondefense sector? Although antitrust is more strictly enforced in the commercial domain, a potentially big step forward was taken when the Microelectronics and Computer Technology Corporation (MCC), a consortium of more than a dozen major U.S. companies, was granted permission to undertake joint research on advanced computer architecture, software, artificial intelligence, component packaging, and CAD/CAM technologies. MCC is, to some extent, an organizational response to the competitive challenge posed by Japanese national projects. Each member company sends researchers to MCC headquarters in Austin, Texas, where they work alongside representatives from other firms. The member companies bear the costs of research and share in its fruits.

MCC is the first consortium of its kind ever to be given the go-ahead to operate as a corporate entity. It performs some of the same functions as Japanese national projects: judicious identification of long-range technological objectives, mobilization of collective resources, research economies of scale, facilitation of information exchange and technological diffusion, and avoidance of wasteful duplication. The Justice Department has warned that it will monitor MCC's activities closely, and it may rescind approval if it uncovers evidence of anticompetitive consequences. Even in the commercial domain, therefore, cooperative research in the United States has been permitted, albeit under stricter constraints and closer monitoring than comparable projects in Japan.

The passage of special development laws in Japan to promote high-tech industries does exempt certain areas from antitrust prosecution. But the exemptions apply only to such matters as standardized specifications for electronics components. Special development laws, in effect for only a limited time, do not give high-tech industries license to engage freely in collaborative research, much less to divide up product and overseas markets, as some charge.[56] The VLSI project, for example, did not qualify for special exemption from antitrust; it fell into the standard category of "laws concerning public enterprise research associations."[57]

Antitrust leniency, based originally on the Law Concerning Temporary Measures for the Promotion of the Electronics Industry (1957), had a major impact on the creation of the Japan Electronic Computer Corporation (JECC) in 1961. JECC is a semigovernmental joint venture comprising Japan's seven leading computer manufacturers; it arranges favorable financing for computer rentals and purchases.[58] It has played a central role in shoring up the competitive power of Japan's computer industry, particularly during the 1960's and 1970's, when IBM threatened to take over Japan's domestic market. Something like JECC would be hard to envision in the United States. It probably would not pass either the Justice Department or the FTC. Here is an instance in which leniency in the application of antitrust law has had long-term consequences.

The overall picture is thus complicated: although antitrust policy in Japan has not permitted Japanese high-tech companies to collude in restraint of trade, it has given them more leeway for collaborative research than U.S. antitrust policy. Moreover, the creation of JECC has had a big impact on the development of the fledgling Japanese computer industry.

Administrative Guidance

One of the most talked-about instruments of Japanese industrial policy, one that has no counterpart in the United States, is administrative guidance (gyōsei shidō)—informal guidelines issued by MITI and other government ministries to help specific industries deal with vexing short-term problems that threaten to harm the collective interest. Unlike formal legislation, administrative guidance is not backed by legal sanctions, since it does not pass through the legislative branch for approval; yet it carries the weight of statutory law in terms of eliciting voluntary compliance. As with other policy decisions, thoroughgoing consultations take place prior to the issuing of administrative guidance. It seldom comes down from above as a unilateral decree, forcefully imposed on a surprised and recalcitrant private sector. Often, private industry approaches the government with the request that certain guidelines be issued.

For MITI, administrative guidance has served as a versatile tool for industrial fine-tuning, particularly during the era of high-speed growth.[59] It has offered MITI the flexibility with which to tailor policies to fit ever-changing circumstances without having to pass a batch of semipermanent laws in the Diet. Over time, the accumulation of such laws tends to clog and constrict the range of policy options. By avoiding heavy dependence on formal legislation, MITI has been able to protect industrial policy from partisan politicking. This has also enabled MITI to intervene selectively and then to pull back. The secret to the effectiveness of administrative guidance is the willingness of the private sector to abide by it, in spite of its nonbinding nature. Such informal guidelines would not work in the United States, where the distance and distrust between government and private enterprise are simply too great to permit coordinated patterns of compliance.

Like other tools of industrial policy, such as antirecession cartels, administrative guidance has come to be used less and less as Japan's economy has matured. For administrative guidance to work, several conditions have usually had to exist: (1) a relatively small number of companies in a given industry that have interacted over a period of time; (2) a clear opinion-leader or market-leader among them; (3) a fairly high degree of market concentration; (4) a mature stage in the industry's life cycle; (5) either a cohesive and strong industrial association or effective mechanisms of industrywide consensus formation; (6) a high degree of dependence on MITI, or at least a history of dependence; (7) common problems of sufficient severity to coax individual companies into cooperating, rather than "cheating," in order to advance collective interests.

Except for the last two, none of the above conditions pertain to the high-tech industries. Consider the software industry. There are literally thousands of companies competing against each other, with no clear leader or strong industrial association, in a dispersed market still in the early phases of growth. Under such conditions it is exceedingly difficult either to arrive at industrywide consensus or to close ranks in pursuit of common objectives. It is not surprising, therefore, that administrative guidance has been used relatively infrequently as a tool of industrial policy for high technology.

One exception is in the area of international trade. Owing to trade conflicts and pressures from foreign trading partners, MITI has had to negotiate a series of binding agreements with domestic companies on matters ranging from dumping to voluntary export restraints. Such agreements have taken the form of administrative guidance with respect to minimum prices, production volume, and export levels. Except for trade-related issues, MITI's reliance on administrative guidance has waned over time.

It is still used for sunset or declining industries that are beset by such structural problems as stagnant or severely fluctuating demand, serious plant overcapacity, and declining international cost-competitiveness. But even for sunset industries, the practice of issuing administrative guidance has been curtailed. In a landmark 1974 decision, the Tokyo High Court ruled that price fixing in the petroleum industry, which had resulted, in part, from private-sector responses to administrative guidance, was illegal. Although stopping short of declaring administrative guidance illegal, the Tokyo High Court noted that limits had to be imposed on its use. Hence, though well known as one of Japan's unique targeting tools, administrative guidance is no longer relied upon extensively, even for old-line industries. Like credit allocation and antirecession cartels, administrative guidance has receded in functional importance over the years, though it continues to be used as a low-key and informal policy instrument.

Industrial Location

Mention should be made, finally, of MITI's role in developing centers of high technology throughout Japan's archipelago. Working closely with local and prefectural governments, MITI has established a network of "technopolises" across the country.[60] The technopolis network is based on the notion that there are advantages to be gained from concentrating facilities for research and development, manufacturing, marketing, and service in compact centers of high technology, like Silicon Valley. The technopolis and companion "science city" provide the advantages of logistical convenience, close communications, close contact, and the synergy that emerges from the existence of a critical mass.

In remote or depressed regions, far removed from the major metropolitan centers (Tokyo-Yokohama, Osaka-Kyoto, and Nagoya), the idea of establishing such centers has struck a responsive chord. It holds forth the prospect of building an enduring infrastructure, ideally suited to the emerging era of high technology; such an infrastructure would produce revenue, employment, prosperity, and demographic balance. From a national standpoint, the technopolis networks would relieve Tokyo and Osaka of the enormous pressures of overcrowding. The desirability of deconcentration and regional relocation has long been recognized in Japan. Former Prime Minister Kakuei Tanaka once proposed a grand scheme for remodeling the Japanese archipelago in a book entitled *Nihon rettō kaizō-ron,* or *A Plan for Restructuring Japan.*[61]

For MITI, the technopolis concept has opened up an avenue for carrying out its responsibility to look after the economic interests of all regions of the country. Industrial location is a central focus of attention in the industrial policies of Great Britain, Italy, France, and West Germany—all countries that have had to grapple with the problem of declining industries in what have come to be known as regional "rust belts." It is also a concern of U.S. officials, even though the federal government has no comprehensive course of action or even a clear center of institutional authority for the development of a long-range plan; the initiative to do something about the regional imbalance has devolved, almost by default, on individual state governments.

In Japan, MITI has retained control over the problem of industrial location; it is an integral aspect of its overall industrial policy approach to high technology. Indeed, one reason for the alacrity with which MITI has grabbed hold of the issue is that it presents an opportunity to extend MITI's influence beyond Tokyo, where it tends to be concentrated, into the farthest reaches of the countryside. To help prefectural governors formulate and implement plans for industrial development, MITI now sends a number of its bright young bureaucrats to serve for short periods as special assistants. This has put MITI in a position to expand its influence in local areas (a position the Construction and Finance ministries have enjoyed for years). It is possible that the establishment of regional influence will serve as a local base for former MITI bureaucrats to run for election to the national Diet, just as the local tax offices serve as a springboard for former MOF officials to run for elective office. For MITI, the extension of its influence into the outlying regions has softened the blow of losing many of the key powers it once possessed.

What impact the technopolis networks and science cities (once they are completed) will have on the competitiveness of Japanese high technology is hard to say. The hope is that they will encourage greater investment; redistribute production, employment, and wealth; construct bigger and better facilities; and create a lasting national infrastructure that will allow Japan to make the transition smoothly from a smokestack to high-technology economy and to sustain itself well into the twenty-first century.

Demand Pull

Technological innovation, production efficiency, and healthy revenues—the sine qua non of competitiveness in high technology—can rest as much on demand pull as on technology push. Although the two are closely connected, some analysts give somewhat greater weight to demand pull.[62] Yet, except in the infant and declining stages of the industrial life cycle, the range and impact of industrial policy instruments for demand pull can be limited, especially if military expenditures are low, as they are in Japan. The available instruments of demand pull include government procurements, home market protection, "buy-national" programs, and export promotion. Yet, as in the case of technology push, the instruments of demand pull, if used unwisely, can distort market forces and cause serious long-term damage. Home market protection and buy-national programs are also capable of giving rise to trade frictions, including violations of GATT-based rules. And the dangers of corruption, fiscal overextension, and politicization loom large because of the sheer size of the public largess.

This suggests that with respect to market demand, the efficacy of industrial policy instruments is limited. Government is poised to make its biggest impact through the macroeconomic stimulation of aggregate demand. Other areas of indirect demand pull that fall into the category of industrial policy include import protection and buy-national campaigns to confine demand to domestic producers. Although such nonmarket measures may bolster the position of domestic producers in the short run, they usually turn out to be self-defeating over the long run because they thwart competition and breed inefficiency.

All sectors have benefited immensely from Japan's postwar growth, the fastest among the large industrial states. High-tech industries have been among the prime beneficiaries because of the high income elasticity of demand for most of their products. One can argue, therefore, that insofar as demand pull is concerned, Japan's high-tech sectors have advanced more as a consequence of sound macroeconomic management than of industrial policies.

Where guaranteed demand as an instrument of industrial policy can register a significant impact is in the early stages of an industry's or a new product's life cycle. The assurance of military procurements, for example, greatly facilitated early innovation in the U.S. microelectronics industry.[63] But the offer of R&D support, by itself, may not be sufficient to persuade companies to assume major investment risks; there must also be the lure of commercial markets. Government-sponsored R&D projects are enticing to some companies only if they are perceived to represent the first wave in what promises to be a second and much larger wave of commercial demand.[64]

Often, the lure of demand pull, combined with the momentum of technology push, is enough to reduce the costs and risks and make forays into the unknown acceptable.[65] The history of innovation in the U.S. semiconductor, computer, and telecommunications industries reveals that the government as guaranteed first purchaser and prime R&D contractor played a crucial initial role, only to be replaced by the subsequent expansion of commercial demand. Government procurement today accounts for only a fraction of aggregate demand in these industries.

MITI: The Lack of Procurements

One of the most striking and significant—but overlooked—features of Japanese industrial policy is the comparative slackness of demand-pull measures. Unlike the U.S. Department of Defense, MITI possesses practically no budget for public procurements. The Japan Defense Agency (JDA), the closest counterpart to DOD, does not make much of a dent on aggregate demand through weapons acquisition. The JDA has one of the world's smallest defense budgets—roughly 1 percent of GNP—of which only about 25 percent is earmarked for weapons acquisition. Of course, because Japan's GNP is so large, the 1 percent figure is not a trivial amount in absolute terms; but even when calculated in yen (rather than percentages), the amount falls far below that of the United States.

The implications of MITI's lack of procurement power are far-reaching. Among other consequences, MITI has had to rely more on a strategy of technology push; for demand pull, high-tech industries have had to rely on macroeconomic measures. Fortunately for them, Japan's domestic market is large enough, and overseas markets are sufficiently open, to allow them to develop without the stimulation of direct demand from the government. But in the absence of an assured customer base of government procurements, Japanese high-tech companies have had to enter consumer-oriented mass markets (such as hand-held calculators for mass memory chips) to recoup the upfront costs of R&D. They established an early foothold in consumer electronics and only recently have moved toward more sophisticated systems applications. The lack of procurement powers has thus forced Japanese companies to compete vigorously in mass consumer markets, a necessity that has caused them to be lean and efficient. The absence of large procurement programs has also spared MITI from excessive political interference, the by-product of corporate lobbying to carve up slices of the public largess.

Whereas MITI lacks the funds for large-scale procurements, other government agencies have substantial purchasing power. The Ministry of Construction presides over a very large budget for public works, and the Ministry of Transportation has drawn on procurement funds to stabilize the ups and downs of shipbuilding demand. There is no doubt that having control over large procurement budgets gives these ministries significant leverage over the private sector; however, such power comes at the price of having to accept a high level of political interference. The ministries of Construction and Transportation, as pointed out in Chapter 4, are among the most politicized bureaucracies in Japan.

Like governments elsewhere, Japanese agencies have tried to bolster domestic companies by purchasing domestically made computers and other high-tech equipment for use in the myriad local, prefectural, and central government offices, public corporations, and public schools. Until 1975, IBM and other foreign manufacturers had been excluded from the circle of companies receiving procurement orders. But in trying to bolster the position of domestic producers, Japan is not different from France or Britain, which gives preferential procurements to national companies. American computer companies in the United Kingdom have gone to court to protest the British government's discriminatory procurement practices.[66] What is noteworthy about Japan is that the level of demand generated by all public procurements is relatively small, accounting for less than 4 percent of the number of computers sold in 1980, and less than 14 percent of the total value of all computers used in Japan.[67]

NTT Procurements

The sole exception to the pattern of limited procurements has been NTT, the public telephone monopoly under the authority of the Ministry of Post and Telecommunications and one of the few public corporations with substantial demand-generating power for the information industries. In 1981, NTT's procurement budget came to about $2.7 billion. For Nippon Electric Corporation (NEC), NTT's largest supplier, procurements exceeded $500 million, representing 12 percent of total company sales. Assured annual sales of telecommunications equipment allowed NTT family firms (especially Hitachi, Fujitsu, NEC, and Oki) to reap valuable benefits, such as economies of scale, greater learning by doing, advances in process technology, large revenues to plow back into R&D, hands-on experience to apply to related technologies, and the capacity to raise more capital.

Until 1981, the procurement bonanza had been closed to foreign bidders; however, under heavy foreign pressure, NTT agreed to open up its procurement system. In 1984, three years after liberalization, foreign manufacturers had managed to win less than 5 percent of NTT procurements, and that mostly in the lower-value-added products, not the higher-value-added areas, like systems integration, in which foreign companies excel. Foreign competitors believe that the closed system bestowed de facto subsidies on Japanese companies; they believe that the competitive ramifications have been felt not only in telecommunications but also in such related fields as semiconductors and information processing.

Demand for telecommunications equipment is expected to keep climbing through the end of the century as NTT implements an ambitious plan to expand and upgrade the nation's entire telecommunications infrastructure through the Information Network System (INS).[68] Over a twenty-year period, INS is expected to generate upward of ¥30 trillion in procurement demand—roughly $230 billion (at the late 1987 exchange rate of 130 yen to the dollar), or an average of over $10 billion per year. If Japanese companies garner all but a small share, the commercial and technological impetus imparted could be enormous for voice recognition and storage, high-speed and mass-volume information processing, and facsimile equipment.

The combination of NTT and JDA demand may facilitate bolder risk-taking and state-of-the-art breakthroughs by Japanese corporations in the future. Large demand stimulation may serve to reduce the risks and costs of new product development and may facilitate innovation in process technology. But defense-related R&D is likely to remain relatively small for the foreseeable future, and the conversion of NTT from a public to a private corporation has transformed the relationship between NTT and its former family firms.[69] The old organizational arrangements for joint R&D have been abandoned, and

NTT procurements have shifted from a closed system of guaranteed demand for NTT family firms to an open system of competitive bidding and price-sensitive market transactions. The change opens up opportunities for foreign companies and international consortia to participate in the booming demand expected to be generated by the growth of Japan's telecommunications industry.

Indirect Demand Stimulation

Without the power of procurements, MITI has had to rely on a variety of indirect instruments to stimulate demand: special tax incentives, export promotion, support for rental and leasing, and home market protection. Until 1983, MITI used special tax incentives to encourage Japanese manufacturers to purchase and install robots, numerically controlled (NC) machine tools, and other automated assembly-line equipment. Although the primary objectives were to raise productivity levels and reduce hazardous working conditions, the switch to automated equipment had the effect of boosting demand for high-tech products. Companies buying robots and NC machine tools received a 13 percent tax credit on the purchasing price, on top of regularly scheduled depreciation allowances.

MITI has also exhorted domestic producers to move down the learning curve by exporting to open overseas markets. Unlike exports of traditional products, like steel, which are sold overseas by giant trading companies, the export of high-tech equipment tends to be handled by the producer companies themselves. Most high-tech products, like computers, require specialized technical knowledge on the part of overseas sales forces and conscientious after-sales service for customers; large trading companies are not particularly well suited to providing the full range of services, from marketing and distribution to customer aftercare. The burden for exporting thus falls squarely on high-tech companies. The Japanese government, through organizations like Japan External Trade Research Organization (JETRO), provides valuable information about overseas markets and sponsors a variety of meetings, conferences, and trade exhibitions. MITI also helps to make arrangements for export financing and insurance through the Japan Export-Import Bank. Since the mid-1980's, when Japan's trade surpluses hit embarrassing highs, MITI has also tried to facilitate industry efforts to find suitable opportunities for direct foreign investment in manufacturing plants abroad.

Lacking the budget to purchase high-tech equipment, MITI has come up with clever ways of accomplishing the next best thing:

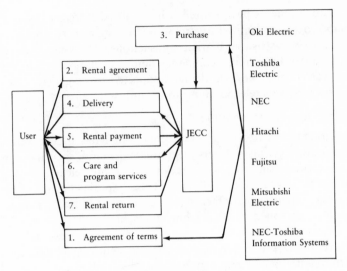

Fig. 2.3. JECC rental system

namely, the organization of nonprofit companies to purchase, rent, or lease costly and continually changing high-tech products. The JECC, created in 1961 as a semigovernmental joint venture involving the seven leading computer manufacturers, is perhaps the best-known example (see Fig. 2.3). Drawing on a pool of funds, including low-interest loans from the Japan Development Bank, JECC had purchased a cumulative total of over $7.25 billion worth of computer equipment by 1981, with rental revenues of $5.6 billion. Its role in helping Japan's computer industry establish a secure foothold in the expanding leasing market was critical, especially during the early years, when IBM appeared to be on the verge of taking control of the Japanese market and Japanese computer firms lacked the resources to compete individually.[70]

The Japan Robot Leasing Company (JAROL), organized in 1980, is another joint venture (comprising 24 robot manufacturers, 10 insurance companies, and 7 general leasing firms) that leases robots to small and medium-sized companies. As of 1982, it had leased nearly 800 robots worth over $25 million. Furthermore, the Small Business Finance Corporation, a government institution, provided loans to small and medium-sized companies for robot installation. Thus, even without a procurement budget, MITI has not been completely handcuffed on demand-pull measures. It has made full use of various indirect policy instruments. While these instruments have not had as much impact as large-scale procurements, they have helped to boost demand in ways that would not otherwise have been possible.

Home Market Protection

Another indirect instrument of demand stimulation on which MITI has relied in the past (but much less since 1980) is the protection of home markets against foreign competition through the imposition of formal import duties, quotas, restrictions on foreign direct investment, and an assortment of nontariff barriers. Formal tariffs, which used to be high as part of Japan's infant industry strategy, have been brought a long way down since the mid-1970's. In 1963 dutiable imports carried a 20.9 percent average levy; by 1983, this had fallen to 4.3 percent. Import duties for high-tech products have also come tumbling down. From 1972 to 1983, computer peripheral duties fell from 22.5 percent to 6 percent; computer mainframe duties fell from 13.5 percent to 4.9 percent; duties on semiconductors dropped from 15 percent to 4.2 percent and all the way down to zero in 1987. Such reductions have brought Japanese duties in line with U.S. levels, and well below those of the European Community.

It can be argued, of course, that it took years of banging on the door before the Japanese government responded and that the barriers came down only after Japanese producers had reached a position of rough parity with foreign manufacturers. Even after liberalization, a variety of nontariff barriers remain, including the unintentional ones of language, business culture, and industrial organization.[71] The thick, tightly integrated network of long-term relationships binding buyers and suppliers, parent corporations and subcontractors, member companies in keiretsu groupings, distributors, financial intermediaries, government, and industry makes Japan an especially difficult market to penetrate (quite apart from other barriers to entry). The difficulties of breaking into the organizational nexus are particularly frustrating for foreign producers of high-tech intermediate goods, because the enclosed and long-term nature of relations between buyers and suppliers alters the character of spot-market, arms-distant transactions.[72]

The barriers to outside entry are also evident in the maze of intercorporate stockholdings that tie Japanese companies together. It may be that the willingness of the Japanese government to accede to foreign pressures to dismantle formal barriers stemmed, in part, from the realization that unintentional structural barriers would still impede full foreign access to Japan's lucrative market. Hiroshi Okumura draws attention to the fact that the impact of capital liberalization, which took place during the late 1960's, has been neutralized by developments within the realm of intercorporate stockholding:

> While the government went about liberalizing capital, Japanese companies were busily implementing their own plans for keeping foreign capital out by means of stock-securing maneuvers. As mentioned, these involved the placement of a large bloc of the stock issued by one company in the hands of other companies—typically its bank and leading partners. The companies thus entrusted with stock did not accept it merely because they were asked to. They became stockholders to enhance intercorporate cohesion and to bind the group together more firmly. The main reason almost no Japanese companies have been absorbed by foreign capital is that each corporate group met the capital liberalization program with its own stock-defense program.[73]

The government encouraged the trend toward intercorporate stock concentration. Loose antitrust enforcement permitted corporate entities to expand their shares of outstanding stocks to over 70 percent, and occasional administrative guidance facilitated the process.

Not only has this obstructed foreign acquisitions and takeovers, an easy path to establishing a physical presence in overseas markets; it has also contained foreign penetration of Japanese markets because of the tight, long-term bonds of structural interdependence between companies in Japan. Buying and selling intermediate goods within a self-contained circle of Japanese companies that own stock in one another makes sense from the standpoint of implicit, long-term contracts and what Oliver Williamson calls "low transaction costs" and what Harvey Liebenstein refers to as "*x*-efficiencies."[74] The logic of business in such a setting is different from that of spot-market transactions. Purchasing intermediate goods from subsidiaries, closely affiliated subcontractors, or members of the same keiretsu can almost be considered quasi-internal transactions. Considering the structure of Japanese home markets, therefore, it is hardly surprising that foreign companies have had difficulty breaking in.

Sales of foreign-made intermediate goods are limited further by the structure of Japanese kaisha as vertically integrated, diversified companies. Instead of meeting all their needs on the merchant market, Japanese computer companies make their own semiconductors and sell what they do not use. Captive, in-house production thus imposes limits on the expansion of foreign shares in Japan's semiconductor market. U.S. semiconductor companies complain that Japanese customers tend to purchase only those highly sophisticated components that they themselves cannot make, such as state-of-the-art microprocessors and custom-made logic devices.[75] Once Japanese companies learn how to produce the advanced devices themselves, the demand for U.S.-made products plummets.

Frustrated American business executives suspect that a well-orchestrated "buy-Japanese" policy lies behind their inability to expand exports. They point out that when they compete with Japanese companies in neutral third markets, such as Europe, they often win the competition and garner larger market shares. But there are other plausible explanations, involving the greater vertical integration among Japanese companies and the self-enclosed nature of Japanese industrial organization. Whatever the reason, the fact is that foreign producers have had a much harder time breaking into, and expanding their shares of, the Japanese market than Japanese producers have had in penetrating the U.S. and European markets.

Japanese electronics companies conduct business within a clear hierarchy of corporate customers, based on the length, magnitude, and overall importance of the business relationship. Priority in terms of financing, services, and supplies is given to corporate customers with which the supplier has had long-standing and extensive relations. Should there be shortfalls of supplies—for example, of 256K DRAMs—Japanese corporations will distribute available supplies in accordance with their hierarchical priorities; companies near the bottom of the hierarchy receive smaller rations. It is not a matter of open bidding or of a neutral, first-come-first-served arrangement. Price is only one factor, and sometimes not the decisive one.

Japanese industrial organization emphasizes the importance of extramarket factors, especially the strong preference for stable, predictable, long-term business relationships based on mutual obligation and trust. This has had the effect of raising the barriers to entry from the outside, whether by foreign manufacturers or by non-mainstream Japanese producers. Such nontariff barriers, deeply embedded in the structure of the industrial economy, are not directly connected to Japanese industrial policy. But their existence, whether by design or by accident, serves basically the same function as formal measures of home market protection—only more effectively, because they do not diminish the vigor of market competition between domestic producers.

Conclusions

There is no alchemy involved in MITI's formula for successful industrial policy. Most of its policies can be found on any standard list compiled by states seeking to promote their high-tech industries. The dosage and mix vary according to specific circumstances and particular needs, but the prescription is fairly standard. MITI's formula can be broken down into three basic elements: technology push, demand pull, and other facilitating measures. Of these, MITI has placed the heaviest emphasis on technology push, featuring national research projects, R&D subsidies, support for basic research, and efforts at technology diffusion. The emphasis comes from the readier accessibility of policy instruments for technology push. It is not derived from a belief that technology push is inherently more efficacious than demand pull. MITI has tried various ways of generating demand, but without a budget for public procurements, its efforts have been limited. Insofar as demand pull is concerned, the Ministry of Finance's macroeconomic measures have registered a far greater impact than MITI's industrial policies.

Nearly all policy instruments that MITI uses in Japan are also employed extensively by governments in other industrial states. As we have seen, Japan is not abnormal in either their adoption or level of use. The only difference is that Japan has somehow managed to wring higher yields from such policy instruments. One reason for the higher yield is that MITI has exercised moderation in its use of all three categories of industrial policy tools—technology push, demand pull, and other facilitating measures. When used to excess, all three categories are capable of doing serious damage to market efficiency.

In coming up with its own formula, MITI has altered the mix of industrial policy instruments that had been applied successfully to the development of old-line industries. It has carefully chosen only those instruments that fit the functional requisites of high technology. Such old-line instruments as structural consolidation and antirecession cartels have been largely abandoned because they are considered too blunt and unwieldy to meet the subtle needs of high technology. The adjustment from smokestacks to high technology is by no means automatic. That Japan has managed to make the transition is a reflection of MITI's capacity to fine-tune market intervention to match the circumstances and goals of specific industries.

Other states have failed to make as smooth a transition. The French government has bought major shares of the stock of high-tech companies like Bull, just as it had in certain areas of banking and heavy manufacturing. State-controlled companies in Italy, like the Istituto per la Riconstruzione Industriale (IRI), hold equity control over key organizations in high technology, just as they do in the old-line sectors. In Britain, the government has entered into venture capital investment, an area restricted to private risk-taking in the United States and Japan. Although the effects of government involvement are not yet clear, the continuing problems of high-tech sectors in Europe suggest, at the very least, that state underwriting is no guarantee of success. Indeed, recent trends toward privatization in England and the sale of government stocks in France may be an admission that extending old industrial policy measures to new sectors of the economy has been a failure.

Selective Omission

Comparing Japanese industrial policy with the policies of other industrial states, one cannot help but notice the absence of certain policy measures that have figured prominently elsewhere. Mention has already been made of the absence of large defense procurements, extensive state ownership of corporate stocks, and state investments in new venture activities—all of which have had costly consequences in terms of economic efficiency in other countries. There are other conscious omissions that have helped Japan to avoid serious economic and political pitfalls.

Although Japanese industrial policy has had the effect of promoting the growth of certain big, blue-chip companies in priority sectors, MITI has not followed a strategy of cultivating one or two "national champions" as France and the small European states have. In large part, as pointed out earlier, this is attributable to differences in size. Japan has a much larger domestic market, population base, and manpower pool, with which it can support more than one or two world-class companies across a spectrum of economic activities. It does not need to concentrate finite resources on the cultivation of a national champion in a few carefully chosen sectors.

If the government attempted to cultivate a national champion for each targeted sector, the effort would violate deeply held norms of fairness and impartiality (not to mention antimonopoly laws); it would also trigger an outburst of outraged opposition. But no such attempt needs to be made. Like the United States, Japan possesses the factor endowments to let a number of strong companies fight it out in the marketplace. The ferocity of the competition strengthens efficiency and the international competitiveness of Japanese firms.

Because it entertains no superpower aspirations in the political-military sphere, Japan has also managed to sidestep another serious pitfall: namely, the disproportionate mobilization of resources for highly visible projects of great pomp and national prestige, but of limited commercial value. Those nations caught up in the "big power" game—the United States, France, and Great Britain—have all succumbed at various times to the temptation of placing national prestige above more mundane considerations of commercial cost-effectiveness. France's program to build a supersonic jet, the Concorde, is an illustration of a high-visibility national project that yielded paltry commercial benefits.

The U.S. Strategic Defense Initiative (SDI) project, the Reagan Administration's ambitious program to build a missile defense system in space, is different from the supersonic jet and airbus projects in its military orientation, but it emerged from the same "big power" syndrome and will probably be as costly in its absorption of finite R&D resources and crowding out of commercial opportunities. Japan has averted such costly mistakes by maintaining a low political-military profile and eschewing big, splashy projects that would thrust it momentarily into the world spotlight. Japan has remained single-minded in its pursuit of only those projects that hold forth the promise of long-run, cost-effective commercial opportunities.

As an ally of the United States, Japan has agreed to abide by the COCOM rules restricting the export of dual-purpose technology to the Communist bloc countries. Except for the Toshiba Milling Machine Company's much-publicized violation in 1987, Japan has adhered fairly faithfully to the letter of the COCOM agreement by not engaging in direct sales to Communist countries. But some U.S. government officials believe that the Japanese have been lax about selling technology to third countries that resell it in turn to the Communist bloc. The U.S. government takes a much tougher stance toward domestic companies trying to circumvent the COCOM guidelines through third-country sales. Indeed, as the military standard-bearer for the Western alliance, the United States administers a restrictive and cumbersome set of controls over the sale of products incorporating sensitive dual-purpose technology of any kind. Chafing under the weight of costly paperwork and frustrating delays, U.S. companies complain that government regulations are handcuffing them in their competition against Japanese firms for lucrative export markets.[76] Here again, the lightness of Japan's military burden looms as a major asset.

Of the selective omissions, however, none is of greater importance than MITI's refusal to cave in to demands that the structurally depressed industries be accorded formal protection against foreign imports. When textile, steel, and automobile producers in the United States could no longer compete, they lobbied the government and persuaded it to pressure foreign producers (especially the Japanese) to accept a series of voluntary export restraints (VERs) that imposed quantitative export ceilings. VER agreements have come to function as vital, life-support systems for many old-line industries in the United States, even though they were originally designed to provide only temporary relief. The problem with VERs is that they erode the GATT framework of trade, delay structural adjustments to the ever-changing international division of labor, and cause serious distortions in the allocation of capital and labor. High-tech industries cannot escape the indirect and multiplier effects of the distortions.

The fact that Japan has not resorted to the imposition of VERs to prop up its declining sectors is of significant indirect benefit to its high-tech industries. It means that they do not have to bear the costs of inflation and inefficiency that semipermanent protection for old-line manufacturers causes. Of course, as pointed out in Chapter 4, the high-tech industries in Japan have incurred costs from the protection of other inefficient sectors, like agriculture; but because steel and car makers are still competitive, Japanese high-tech companies have a healthy set of downstream industries on which to rely for the consumption of their intermediate goods. In an economy of complex interdependence, the benefits of efficiency can be passed along, just as the costs of inefficiency are bound to be transmitted in the form of indirect multiplier effects.

From a long-term, economy-wide perspective, the United States' imposition of a succession of VERs on Japan has had a boomerang effect. It has pushed Japan steadily up the ladder of higher value added in the composition of its exports to the United States. When VERs were placed on textiles, Japan stepped up its sale of steel; when VERs were placed on steel, Japan expanded its market shares of cars and consumer electronics; and when VERs were placed on cars and consumer electronics, Japan shifted to the export of high-technology products. Throughout, total export volume has continued to rise. By clamping down on one product category (old-line manufacturing), the United States merely accelerated Japan's speed in moving into the production and export of more sophisticated and expensive goods (high technology). In a very real sense, therefore, U.S. protectionism in the declining, old-line sectors has had the perverse effect of making the Japanese more competitive in high technology.

Although Japan has not forced other countries to swallow VERs, textile producers in China and steel and automobile companies in Korea have complained about being shut out of Japanese markets. They believe Japan is relying on hidden nontariff barriers instead of VERs to protect domestic producers in old-line sectors. Whether and to what extent this allegation is true is hard to determine. If it is true, and if the scope of protection is as far-reaching as they claim, then Japan is apt to suffer the same boomerang effects as the United States.

There is, however, a significant difference between Japan and the United States: namely, that the Japanese government has coordinated comprehensive plans to cut back plant capacity in the structurally depressed sectors. Nearly 50 percent of total plant capacity has been scrapped in old-line industries like shipbuilding, aluminum smelting, certain areas of textile production, and petrochemicals. Steel and other old-line industries are similarly undergoing structural reductions and automated upgrading. Moreover, with the yen's sharp upward revaluation, Japanese old-line manufacturers are investing in offshore production facilities through the vehicle of joint ventures with indigenous partners. Seeing the handwriting on the wall, many are also diversifying into higher-growth fields; a number of textile companies, for example, have entered into the field of biotechnology.[77] Through "positive adjustments," therefore, Japan is relieving some of the pressures that would otherwise mount for the protection of inefficient old-line industries. Here, as in other respects, Japan bears a striking resemblance to the small European states, which feel they have no choice but to bow to the forces of the international economy.[78]

Assessment

Although most instruments of Japanese industrial policy are neither unique nor abnormal, a few take distinctive form in Japan. Although they serve the same functions as policy instruments in other industrial states, they have no precise equivalent elsewhere. Administrative guidance, special temporary development laws for promoting priority industries, the old NTT family setup, and positive adjustment policies for the scrapping of excess plant capacity fall into this category. All have contributed something to the overall effectiveness of Japanese industrial policy. The old NTT family system for cooperative research and procurement and positive adjustment policies designed to streamline surplus plant capacity have been especially constructive and distinctive policy instruments.

Japan has also had some measure of success, where others have experienced nothing but failure, in the organization of national research projects, the encouragement of both competition and cooperation, the insulation of industrial policy from politicized forces, the limitation of damage by the distortion of market forces, the comparative coherence of MITI's industrial policy, and the relative absence of irreconcilable conflicts between industrial and macroeconomic policies. These represent considerable achievements when compared to industrial policy in most other countries. Being able to keep industrial policy from falling into the hands of politicians, for example, is a rare feat; the same holds true for all the other achievements listed above. Few other countries have managed to replicate Japan's accomplishments, in spite of using most of the same instruments of industrial policy.

Why is it that countries using basically the same set of industrial policy instruments should experience such divergent results? The question suggests that the policy measures themselves are not the sole factors determining the success or failure of industrial policy. Although unwise, market-defying measures will surely subvert economic efficiency, market-conforming measures will not always produce successful results. In other words, the selection of sound measures may be a necessary but not sufficient condition of an effective industrial policy. Effectiveness depends on other factors, especially the characteristics of market organizations and the structure of political institutions.

In Japan's case, the relative efficacy of the policy instruments used for the promotion of high technology has depended on the distinctive nature and structure of Japan's political economy. Sound instruments were chosen and then made to work because of the existence of such systemic factors as the consensual mode of policy-making, the extensive information gathering and analysis on which consensus is based, and the tradition of close government-business relations—all of which emerge from the strengths of Japan's political economy. To understand why industrial policy has not created more problems than it has solved in Japan—when it has given rise to a multitude of problems in countries like Italy and Britain—requires that we move the focus of our analysis beyond the repertoire of policy instruments. At a higher and more abstract level, we must analyze the regime characteristics of Japan's political economy, especially the organization of its private sector and its relationship to the institutions and processes of the political system. It is to an analysis of these subjects that we turn in Chapters 3 and 4.

THREE

MITI and Industrial Organization

The preceding chapter alluded to the importance of industrial organization in structuring the patterns of interaction between government institutions, like MITI, and private corporations. The government's choice of certain industrial policy measures and its capacity to implement them effectively often depend on such aspects of industrial organization as capital and labor markets. In Japan's case, MITI's capacity to administer industrial policy for high technology has hinged on the organizational characteristics of Japan's industrial and political system.

It is to an analysis of Japanese industrial organization and relations between MITI and business that we now turn. We shall begin with a brief analysis of MITI, focusing on the organizational characteristics that help to make it effective.[1] We shall then examine features of Japanese industrial organization, calling attention to those that explain why private-sector interactions with MITI have been constructive. The chapter will conclude with a look at the role played by intermediate organizations standing between the public and private sectors, particularly the informal, ad hoc policy networks linking MITI bureaucrats with businessmen.

MITI: The Crucible

MITI is almost always cited as one of the main reasons for the effectiveness of Japanese industrial policy.[2] Among large industrial states, few, if any, bureaucracies exercise comparable power over the sector-specific management of the industrial economy. For a variety of reasons (discussed in Chapter 4), the ruling Liberal-Democratic Party (LDP) has permitted MITI unusual leeway to oversee the operation of the country's manufacturing and service sectors. MITI's autonomy also has its roots in the distinctive circumstances surrounding Japan's postwar history, especially the extraordinary powers over foreign exchange, foreign direct investment, and technology licensing conferred by the Allied Occupation. From the end of the war to 1980, MITI served as the linchpin for a system of latecomer development, perhaps the most effective of its kind ever devised. Once in place, this latecomer system, which Chalmers Johnson has called the "developmental state," routinized the processes of industrial policy-making.[3]

Another reason for MITI's power is the extraordinary quality of its higher civil servants. Every year, MITI hires from 25 to 30 of Japan's "best and brightest," most of whom hail from the University of Tokyo's Faculties of Law and Economics. Competition for the crème de la crème of the college crop is especially strong between MITI and the Ministry of Finance because the two bureaucracies are considered the most prestigious in Japan for those who have passed the grueling Higher Civil Service Examination. What makes the two ministries so attractive is not only the power and prestige they command but also the availability of lucrative second-career jobs (*amakudari*) in the private sector when the time comes for higher civil servants to retire. In the competitive recruitment derby, MITI has managed to hold its own against other ministries, including MOF. The quality of its career officials is second to none.

Superior personnel constitute, without question, one reason for MITI's enviable performance; but they are not the sole explanation. Civil servants in several European states, notably France and England, also come from elite backgrounds;[4] yet none of the European bureaucracies can boast track records that match MITI's. What makes MITI stand out is the effectiveness of the industrial and political system within which it operates. It would be difficult to imagine a MITI-type organization functioning as smoothly in the United States, France, or England, where it would not be reinforced by a complementary set of institutions.

Placing MITI in the context of Japan's overall political economy, however, should not diminish the significance of its own formidable strengths. The comprehensiveness of its administrative jurisdiction and the cohesiveness of its internal organization have to be considered important reasons for its effectiveness. MITI's mandate, to begin with, is extraordinarily broad. As pointed out in Chapter 1, MITI has charge of everything from energy security to retail distribution, textiles to aircraft, heavy manufacturing to leisure activities, and regional development to international trade and investment. The companies it deals with come in all sizes and shapes, ranging from tiny family firms that operate out of small cottages to mammoth trading companies with sprawling office complexes scattered all over the globe. Only a handful of key sectors, including the primary industries (agriculture and fishing), health-related services (medicine and health care), shipbuilding, domestic construction, financial services, and aspects of telecommunications fall outside its jurisdiction. Having authority over so wide a swath of Japan's economy is an asset of enormous value for MITI.

The rational way MITI is organized has permitted it to parlay this near-monopoly of authority into an effective tool for industrial policy. When administrative authority over the industrial economy and foreign trade is badly fragmented, as it is in the United States—where it is shared among the Departments of Commerce, Defense, State, and Energy, the United States Special Trade Representative (USTR), Congress, and the White House—a coherent industrial policy is virtually impossible to achieve.[5] If MITI's administrative responsibilities were less sweeping—if, for example, the Ministry of Post and Telecommunications had a broader mandate, if the Japan Defense Agency had greater clout, or if separate ministries existed for energy, small and medium-sized enterprises, and industrial science and technology—the likelihood is that Japanese industrial policy for high technology would also be more fragmented and therefore less effective.

Thanks to its broad jurisdiction over most of the key sectors of Japan's industrial economy, MITI has been able to take a general equilibrium approach to industrial policy, weighing the interests of one sector against those of others and aggregating the welter of diverse, sometimes conflicting, demands into a comparatively coherent whole. Chapter 2 discussed MITI's internal system of allocating subsidies and special tax incentives for priority industries within set ceilings. Not only is this system conducive to the maintenance of fiscal discipline, it also protects MITI from being held hostage by special-interest groups. No single industry or interest group, no matter how powerful, exercises dominant influence on MITI or dictates the substance of industrial policy.

When companies in sunset sectors lobby for import protection, their demands collide with MITI's commitment to positive adjustment policies (PAP) designed to adapt Japan's industrial structure to shifting comparative advantage. Protectionist demands run squarely against the interests of healthy sectors, which would have to bear the indirect costs of inefficiencies caused by protectionism. Within MITI, the short-term demands of noncompetitive industries for temporary relief from foreign competition have to be weighed against the long-term goal of maintaining an efficient and dynamic economy. The process of interest aggregation that takes place within MITI contrasts sharply with the fragmented U.S. system, in which organized industrial and labor interests always seem to find ways of infiltrating the institutions of policy-making, especially Congress. Special concessions such as tax breaks for declining industries are not cost-free; other industries and individual taxpayers wind up paying the price.

MITI's organization into vertical and horizontal units makes interest aggregation less costly in Japan. Vertical units, like the Machinery and Information Industries Bureau (Kikai Jōhō Sangyō-kyoku), deal with problems in specific industries, while horizontal units, like the International Trade Policy Bureau (Sangyō Seisaku-kyoku), handle functional issues cutting across sectors (see Figs. 3.1 and 3.2). This organizational structure creates a system of internal checks and balances: the tendency of vertical units to push for support for industries under their supervision is balanced by the broader vision of the functional horizontal units, which look out for the interests of the industrial economy as a whole. *Genkyoku* officials (those responsible for specific industries) can state their case for special policy measures before colleagues in the functional units; all sectors are assured of a fair hearing. Functional units like the General Coordination Division of the Minister's Secretariat are responsible for aggregating specific interests across all sectors of the industrial economy under MITI's supervision; they cannot allow individual sectors—no matter how powerful—to dominate the policy-making processes or dictate the content of industrial policy.

MITI's capacity to aggregate diverse interests is further safeguarded by the structure of interdivisional councils, which incorporate both vertical and horizontal bureaus.[6] There are three such policy-making councils, one at each of the following levels of seniority: deputy division directors (*kachō hosa*), division directors (*kachō*), and directors general (*kyokuchō*). The most influential is the lowliest in terms of age and seniority, the Hōrei Shinsa Iin (Policy Legislation Deliberation Council), which comprises deputy directors from the General Affairs divisions of each MITI bureau, plus their counterparts from the General Coordination, Budget and Accounts, and Personnel divisions of the Minister's Secretariat. The Hōrei Shinsa Iin is the key interdivisional organ for policy formulation within MITI. It brings together young deputy division directors, who function as the most active

	Vertical bureaus MITI: manufacturing, commercial, service industries					Other ministries		
Horizontal bureaus	Basic Industries Bureau (iron and steel, nonferrous metals, chemicals, biotechnology, etc.)	Machinery and Information Bureau (industrial electronics, consumer electronics, data processing, aircraft, automobiles, industrial machinery, cast and wrought products, space, etc.)	Consumer Goods Industries Bureau (fibers and spinning, textiles, paper and pulp, household miscellaneous goods, ceramics and construction materials, housing, traditional crafts, etc.)	Agency of Natural Resources and Energy (electrical power, gas, petroleum, coal, nuclear energy, ocean development, etc.)	Industrial Policy Bureau (commerce, service trade)	Agriculture	Construction; Transportation; Telecommunications	Finance (banking, securities, insurance)
Industrial policy: Minister's Secretariat; Industrial Policy Bureau								
Natural resources and energy policy: Agency of Natural Resources and Energy								
Small-business policy: Agency for Small and Medium-Sized Enterprise								
Patents: Patent Office								
Government Industrial Technology: Agency of Industry and Development								

Fig. 3.1. MITI: Vertical and horizontal organization

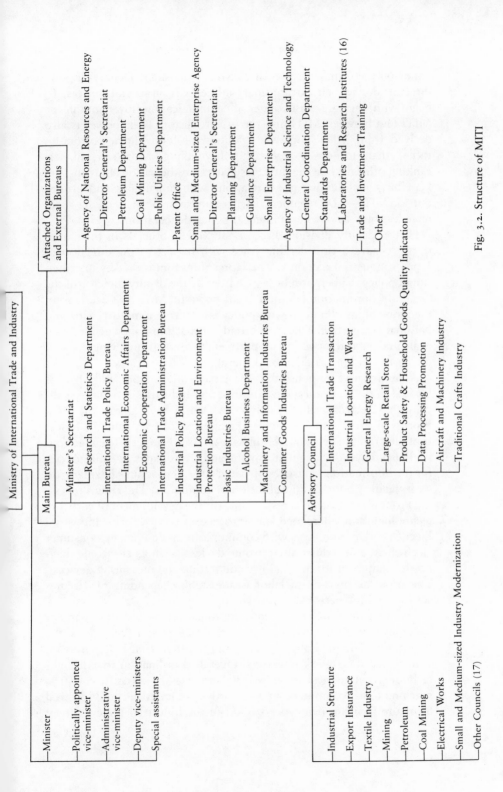

Ministry of International Trade and Industry

Minister
Politically appointed vice-minister
Administrative vice-minister
Deputy vice-ministers
Special assistants

Main Bureau
Minister's Secretariat
Research and Statistics Department
International Trade Policy Bureau
International Economic Affairs Department
Economic Cooperation Department
International Trade Administration Bureau
Industrial Policy Bureau
Industrial Location and Environment Protection Bureau
Basic Industries Bureau
Alcohol Business Department
Machinery and Information Industries Bureau
Consumer Goods Industries Bureau

Attached Organizations and External Bureaus
Agency of National Resources and Energy
Director General's Secretariat
Petroleum Department
Coal Mining Department
Public Utilities Department
Patent Office
Small and Medium-sized Enterprise Agency
Director General's Secretariat
Planning Department
Guidance Department
Small Enterprise Department
Agency of Industrial Science and Technology
General Coordination Department
Standards Department
Laboratories and Research Institutes (16)
Trade and Investment Training
Other

Advisory Council
International Trade Transaction
Industrial Location and Water
General Energy Research
Large-scale Retail Store
Product Safety & Household Goods Quality Indication
Data Processing Promotion
Aircraft and Machinery Industry
Traditional Crafts Industry

Industrial Structure
Export Insurance
Textile Industry
Mining
Petroleum
Coal Mining
Electrical Works
Small and Medium-sized Industry Modernization
Other Councils (17)

Fig. 3.2. Structure of MITI

hands-on policymakers in each of MITI's divisions. Deputy division directors derive their extraordinary influence from several sources. To begin with, they are at the center of communication flows both inside MITI and between MITI and the private sector. Their power is also derived from MITI's postwar tradition of following a bottom-up rather than top-down style of policymaking. Moreover, higher-ranking officials, like the directors general of the various bureaus, are kept busy maintaining good relations with individuals and groups outside MITI—LDP members, Diet committees, the mass media—and have less time to devote to internal policy deliberations.

Hence, deputy division directors at MITI—officials in their mid-thirties—have a considerable voice with respect to the direction of Japan's industrial economy. The Hōrei Shinsa Iin is the key interdivisional policy deliberation body. Chaired by the deputy director of the General Coordination Division of the powerful Minister's Secretariat, the Hōrei Shinsa Iin deals with policy issues from a ministrywide and national perspective. The same broad approach to policy deliberation is also taken at the two interdepartmental councils involving division directors and bureau directors general.

Administrative responsibility for Japan's information industries is located in the Machinery and Information Industries Bureau, specifically the Electronics Policy Division (Denshi Seisaku-ka), the Data Processing Promotion Division (Jōhō Shori Shinkō-ka), and the Industrial Electronics Division (Denshi Kiki-ka) (see Fig. 3.3). The Electronics Policy Division formulates plans and coordinates overall policies for the information industries. It is in charge of programs having to do with the distribution and utilization of computers. The Data Processing Promotion Division is responsible for the development of software technology and all related data-processing services. The Industrial Electronics Division deals with communications equipment, consumer electronics, and industrial electronic devices (such as electronic diagnostic equipment for medical use and testing and measuring devices). The tripartite division of labor is adjustable, depending on circumstances and the personalities of key policymakers.

The three divisions stay in constant touch with many other parts of MITI, including the Agency for Industrial Science and Technology, the International Trade Policy Bureau, and the Industrial Policy Bureau, with which their work intersects. Overall coordination for all high-technology industries under MITI's supervision—aircraft, robotics, machine tools, computers, and so forth—is located in the Industrial Structure Division (Sangyō kōzō-ka) of the Industrial Policy Bureau

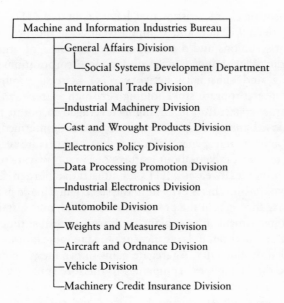

Fig. 3.3. Structure of the Machine and Information Industries Bureau of MITI

(Sangyō Seisaku-kyoku); for matters related to high technology financing, responsibility rests with the Industrial Finance Division (Sangyō shikin-ka) of the Industrial Policy Bureau. Prior to 1984, there was no clearly designated center for overall coordination of high-technology policies. Although the three genkyoku divisions under the Machinery and Information Industries Bureau still retain primary responsibility for the information industries, their policy proposals have to fit into an overall program for high technology.

MITI and the Private Sector

MITI does not operate in a vacuum. No matter how rational its internal organization or how capable its career bureaucrats, MITI's effectiveness ultimately depends on its capacity to work in harmony with the private sector. Indeed, a case can be made that MITI's track record owes at least as much to the effectiveness of its interactions with the private sector (made possible by the nature of Japanese industrial organization) as it does to the strengths of MITI's internal structure. Certainly the task of administering industrial policy would be far more burdensome were it not for the dynamism of the private sector and the distinctive features of Japanese industrial organization.

Private-sector characteristics that have facilitated MITI's tasks include (1) the efficiency and competitiveness of the private sector; (2) enterprise unions and the comparative weakness of organized labor; (3) the single-minded focus of Japanese corporations; (4) subcontracting and subsidiary networks; (5) keiretsu groupings; (6) extensive intercorporate stockholding; (7) high keiretsu-adjusted ratios of market concentration; (8) highly leveraged corporate financing; (9) very close banking-business relations; and (10) intermediate organizations and the interpenetration of public and private sectors.

One of the most talked-about features of Japan's private sector, and one that dramatically eases MITI's administrative burden, is its competitive dynamism. This is evident in the steep gains made in industrial productivity throughout the postwar period. High and sustained rates of capital investment in new plant facilities—the substitution of capital for labor—account in large measure for Japanese improvements in value-added productivity. Aggregate gains in productive efficiency, in turn, have thrust Japanese companies to the forefront of world competition in manufactured goods.

Like the steel industry during the 1950's and 1960's, Japan's information industry has found itself locked into an investment race in the 1980's. Except for commodity chips, however, this has not led to the dysfunction of excess plant capacity as it did in steel because market demand is still on an upward trajectory. Moreover, the leeway for product diversification, such as increased demand for ASICs, alleviates the kind of pressure that can lead to excessive competition. So far at least, MITI has not had to worry much about neutralizing the downside effects of the investment scramble (except for DRAMs). It has enjoyed the luxury of letting Japan's information industries fight it out in the marketplace. The competition has spurred aggregate growth rates, the pace of technological innovation, manufacturing efficiency, price reductions, quality improvements, product differentiation, and the expansion of marketing and after-sales service. Nothing makes MITI's job easier than the dynamism of vigorous, self-sustaining competition in the private sector.

Competition-driven efficiency in certain areas of old-line manufacturing has also relieved MITI of the otherwise onerous burden of designing industrial policy that would overcome the loss of comparative advantage. In contrast to the experiences of governments in Italy, England, and France, MITI has not kept dying industries on life-support systems by nationalizing, subsidizing, and protecting ailing companies.[7] The private sector's capacity to compete with foreign producers permits MITI to take a less interventionist and less costly stance.

Japan has also not had to adopt the tactic used by most European states: namely, to resign itself to specialized niches of international markets. Nor has it had to fall back on the promotion of foreign joint ventures as second-best substitutes for the ability to compete in state-of-the-art technology. It has had the enviable freedom, instead, to devise policies that supplement, not substitute for, market forces. Having a vigorous private sector thus gives MITI flexibility in designing industrial policies that allow it to refrain from excessive market intervention.

The Muted Voice of Labor

One of the most noteworthy features of the Japanese political economy, one that sets it apart from virtually all Western European countries, is the weakness of organized labor. In Japan, labor is vertically organized into so-called enterprise unions within individual firms, not as craft unions across companies and along functional or horizontal lines. On the shop floor and inside corporations, Japanese labor is given a fairly strong voice, thanks to the system of consensus decision making. But at the level of central government—inside the corridors of the bureaucracy and parliament—Japanese labor does not exert nearly as much influence as organized labor in Western Europe.

Inside companies, Japanese unions do not have anything comparable to the system of contractural rights accorded to workers or to worker representation on corporate boards, as many West German unions do. This is a reflection of the fact that Japanese labor has less formal power than labor unions in West Germany, Austria, Norway, or Sweden. However, under Japan's consensual system of decision making, labor unions do have considerable say in matters of direct and tangible concern to rank-and-file members, such as salary increases, employee benefits, working conditions, personnel matters, and assembly and process operations. They have much less say in more distant decisions concerning product portfolios, R&D investments, and corporate strategy. Hence, although Japanese management has had to respond to labor demands in specific areas of workers' concerns, it has enjoyed a comparatively free hand in matters related to corporate strategy.

The capacity of labor and management to deal with each other across the industrial divide has led to remarkably low levels of strife.

Both sides have benefited enormously from the harmony: Japanese workers have seen their real incomes rise to levels common in the United States, while management has presided over the ascendance of Japanese corporations as world-class competitors. The low level of industrial conflict is reflected in the low number of days lost to strikes and disputes: 252 days were lost in Japan in 1986 compared to 568 in France, 12,140 in the United States, and 1,879 in the United Kingdom.[8] The self-regulating nature of labor-management relations has spared the Japanese government from being engulfed by the consuming task of binding up economic and social wounds following outbursts of labor unrest. It has given both MITI and Japanese management enviable leeway to maneuver. MITI has not had to participate in the kind of neo-corporatist bargaining that takes place among government, business, and labor in Western Europe.

MITI has also benefited from the existence of a de facto incomes policy that functions on a self-regulating basis within the private sector.[9] It is not an incomes policy set by an act of government. Nor is it the by-product of neocorporatist institutions that negotiate binding agreements among industry, labor, and government, as is the case in Austria and Scandinavia. Rather, Japan's de facto incomes policy has grown organically out of a highly routinized set of norms, procedures, and institutions developed over years of interaction between labor and management.[10] Wage bargaining in Japan leaves little to chance and provides both labor and management with a highly predictable framework within which to arrive at satisfactory settlements.

The whole process is based on extensive preparatory spadework and prior consultations (*nemawashi*), negotiations over a fairly narrow band of wage figures, early agreement by a few leading companies and industries, the establishment thereby of a general standard or norm, and the rapid diffusion of that norm to all sectors. Besides labor and management, prominent leaders from academia, government, research institutes, industrial associations, and the mass media take part in the process. The negotiating process unfolds almost as if it followed a pre-ordained script, free of unexpected twists, government jawboning, or prolonged, paralyzing strikes.

During the era of high growth, especially from 1965 to 1974, Japanese labor unions obtained generous wage hikes. Since 1975, however, under much harsher business conditions, they have accepted lower wage settlements in exchange for greater job security. The enterprise nature of Japanese labor unions has disposed them to take a less confrontational approach toward management than their counterparts in the West, since the security of their jobs depends on the survival and prosperity of the companies for which they work.

From a macroeconomic perspective, Japan's de facto incomes policy has had the effect of alleviating the pressures of cost-push inflation, providing some relief from the struggle against the vexing two-headed problem of inflation and unemployment. From the standpoint of industrial policy, the informal system of wage restraint has had the effect of holding down the labor component of production costs, defusing potential labor-management discord, and promoting equity in income distribution; the first helps to keep Japanese companies internationally competitive, the second minimizes the loss of valuable work time, and the third fosters a sense of social justice. This has lightened considerably the public policy work load that would otherwise have fallen on the government's shoulders. As in the case of private-sector competitiveness, MITI has not had to face some of the difficult problems engendered by labor-related conflicts.

At the national level of electoral politics, the voice of organized labor is not nearly as strong as it is in most Western political economies. In contrast to Sweden, France, and England, labor-based political parties in Japan not only have failed to win majorities in the parliament but have also been completely excluded from participation in the succession of cabinets formed in the postwar period (save during a short interlude in the late 1940's). As for the formulation of economic policy, Japan's labor-based parties have been relegated to playing a peripheral role; the ruling conservative party and the bureaucracy have worked in tandem to steer the economy in directions tending to favor producer interests. This has posed a stark contrast to the pivotal role played by labor-based parties in the creation of Western European welfare states.

This is not to say that the socialist parties have had absolutely no impact on the formulation of economic policy; they are assured at least some input by virtue of the norms of parliamentary procedure requiring that they be consulted and allowed to express their opinions. And beyond the routine consultations associated with parliamentary etiquette, socialist parties can exercise some leverage by picking their targets carefully and threatening all-out opposition to specific pieces of objectionable legislation. The threat of an all-out floor fight carries significant weight, especially if all opposition parties join together to form a united front. Under such circumstances, the LDP, which possesses sufficient numbers, can resort to forceful passage (*kyōkō saiketsu*) over the strenuous objections of a unified opposition, but this would violate the norms of acceptable parliamentary behavior. The LDP would come across to the public as high-handed and authoritarian. In this sense, the labor-based parties have managed to exert a real measure of veto power.

In 1987, the labor-based parties combined with other opposition parties to mount a determined effort to block Prime Minister Yasuhiro Nakasone's plans to pass a value-added tax. Although Nakasone had promised during the 1986 election campaign not to impose new taxes, he went back on his pledge and tried to push through the Finance Ministry's value-added tax plan, drawing confidence from the mandate he felt the landslide electoral victory had given him. The opposition parties fought Nakasone's tax reform, and various interest groups, especially owners of small and medium-sized companies (whose books would have to be opened to closer scrutiny), lobbied hard to block the value-added tax. In the end, Nakasone was forced to withdraw his plan. For him and the LDP, it was a humiliating setback, possibly the worst suffered during his years as prime minister; but it demonstrated the capacity of opposition forces, centering on labor-based parties, to exercise the negative power of the veto.

The death of the value-added tax plan, however, cannot be considered a typical example of the labor-based parties' actual and latent power. The veto effort probably would have ended in failure if politically powerful interest groups like small and medium-sized enterprises had not fought so hard. It certainly would have failed if opposition parties had not joined together. On most issues, the opposition forces have shown a chronic inability to transcend their petty bickering and forge a united front. Without a unified front and strong interest-group backing, the labor-based socialist parties cannot exercise much influence, since they lack the leverage that comes with the threat of legislative blockage. Under normal conditions, therefore, Japan's bureaucracies and the LDP enjoy an unusual degree of freedom from organized labor and labor-based parties in formulating industrial policy. The Japanese government's extraordinary degree of freedom from the influence of organized labor can be seen in such areas as the reduction of surplus plant capacity in structurally depressed industries.

Company Specialization

Another noteworthy feature of Japanese private enterprise is the conspicuous tendency of most corporations in growing sectors to stick to their main line products, the "core businesses" on which they were established.[11] Leaving aside declining or stagnant sectors, successful Japanese companies tend to be wary of diversifying into unrelated fields, or acquiring companies for financial or tax reasons, or taking major steps to protect themselves against hostile takeovers. They are even less inclined to merge with companies in totally different areas.[12] Except for a small handful of diversified giants, like Mitsubishi Heavy Industries and Hitachi, the majority of Japanese companies prefer to stay within the confines of a single, clearly defined industry.

This preference can be explained largely in terms of labor market conditions. Because large Japanese corporations practice career-long employment, it is not easy for them to move into new and wholly unrelated fields. Horizontal diversification may require that they recruit a new cadre of workers. It may mean taking on a major, long-term commitment of resources, not to mention venturing into unexplored terrain. Since Japanese firms cannot lay off employees as easily as U.S. companies, significant departures from the core business involve substantially higher risks and costlier commitments than is the case in the United States. Mergers and acquisitions, alternatives to internal diversification, pose even more troublesome problems, since companies must deal with the headaches of trying to integrate two separate work forces. Lifetime employment thus forces Japanese companies to be cautious about extending themselves into areas where they possess no prior experience, related expertise, or comparative advantage; on the other hand, the commitment to career-long employment is prompting companies in declining industries like steel to diversify.

Labor market practices also explain why Japanese corporations prefer to set up their own networks of independent subsidiaries instead of expanding internally. Creating subsidiaries can be an appealing alternative to entry into new markets through vertical integration, backward or forward.[13] No risky, long-term commitments, which overextend the fixed or slack resources of the parent company, need be made. The parent firm can concentrate on what it does best while benefiting from what the new subsidiary is set up to do. For highly leveraged Japanese companies, the pitfalls of diversification—cyclical vulnerabilities and the rigidity of higher fixed costs—can be circumvented. If the subsidiary is an upstream operation, suitable subcontracting arrangements with the downstream parent company can be worked out. If the subsidiary is a downstream operation, it can benefit from the parent company's upstream capabilities. Moreover, with subsidiaries, parent companies have the leeway of reassigning surplus employees, an important safety valve for the cumulative pressures of permanent employment.

From the standpoint of implementing industrial policy, the pattern of self-contained specialization is of significance because it simplifies MITI-industry interactions. MITI officials can concentrate on developing enduring relations with a well-defined and delimited set of companies in each industry. They do not have to deal with a maze of divisions in conglomerate hierarchies. It would be harder to deal with multidivisional units in a conglomerate than with independent companies in a single industry, because the particular interests of the multidivisional units might get overwhelmed. This would make it difficult to build a binding sense of industrial community or to define a common set of interests and goals. It might even be hard to define what constitutes an industry, since company boundaries would not be clearly demarcated. Under such circumstances, MITI probably would have trouble tailoring industrial policy to meet the special needs of each industry.

MITI's relationship with general trading companies (*sōgō shōsha*) illustrates the difficulties involved in dealing with highly diversified giant corporations. One would expect the MITI–trading company relationship to be very close, given the latter's size, market power, economic impact, and dominant role in domestic and international trade. The two sides have so much in common that extensive cooperation would seem almost a foregone conclusion. But such is not the case. There is surprisingly little cooperation and not even much routine interaction, except on such mundane matters as export-import insurance.

On industrial policy issues, the general trading companies are scarcely consulted. The arm's-length relationship is due, in part, to the low cost-benefit yield of dealing with giant conglomerates, which, because of their size and diversity, cannot deal quickly or flexibly with the specific needs of individual industries. Another reason lies in the power of general trading companies, which are not as dependent on MITI as most manufacturing concerns and therefore not as amenable to cooperation (particularly when it entails sacrifices to corporate interests).

In addition to company specialization, Japanese industrial organization is also characterized by the small presence and comparative insignificance of foreign manufacturers in Japan. Nearly all high-priority industries are dominated by Japanese producers; only a handful of foreign subsidiaries and foreign-majority-owned joint ventures are represented. The computer industry is one of the few in which a wholly owned foreign subsidiary—IBM Japan—is a major actor on the domestic scene. The small share of key Japanese markets held by foreign manufacturers is due in part to the characteristics of Japanese industrial organization and in part to the strict controls over foreign direct investment that made it difficult to establish a strong foothold in Japan until liberalization in the late 1970's.

With so few foreign-owned companies to deal with, MITI and Japanese industries have had a free hand in formulating and implementing national industrial policy. They have not had to face embarrassing questions or strong objections from locally based foreign manufacturers; few foreign-owned companies are represented on the membership rosters of Japanese industrial associations or on blue-ribbon advisory panels. Japan's exclusive old-boy system has made it easier for MITI and private companies to discuss policy issues uninhibitedly, abide by unspoken rules and norms, implement certain kinds of policies (like administrative guidance and antirecession cartels), and pursue national goals and collective interests. It has lowered the transaction costs of collective action. Whether Japanese industrial policies would have been well coordinated without this homogeneity is doubtful. Such clubbishness, however, has given rise to foreign complaints about the lack of transparency in the industrial policy-making processes.

The tidy way each industry is organized facilitates enormously MITI's capacity to carry out all the functional tasks associated with industrial policy—staying in touch with the major companies in each industry, defining a vision of the industry's future, building industry-wide consensus, and mobilizing the resources necessary to achieve collective goals. Industrial associations can be organized within reasonably clear functional boundaries, thanks to the strong inclination of Japanese corporations to stay within the perimeters of their core businesses. MITI officials find themselves in an ideal position to develop informal policy networks with a manageable core of industrial leaders.

Being able to fulfill industry-specific tasks at acceptable transaction costs is crucial to MITI's effectiveness, to the maintenance of healthy MITI-industry relations, and to the success of Japanese industrial policy. If Japanese companies strayed as often and as far from the boundaries of their original, core businesses as U.S. companies do, designing and implementing an intelligent set of industrial policies would certainly be more difficult. Here is an example of fortuitous convergence: the institutional structure of private enterprise—which has come into existence largely as a by-product of permanent employment—provides an ideal administrative framework for industrial policy-making.

It should be pointed out, however, that although the structure of Japanese industry is more compartmentalized than that in other large industrial states, like the United States and West Germany, the degree of company diversification varies by sector, and the trend is toward greater diversification. Here again, there seem to be fundamental differences between old-line and high-technology sectors. A number of old-line industries that have lost or are in the process of losing comparative advantage in original product markets, like textiles, have begun to diversify into new but related lines of business. Many of the new entrants in the field of biotechnology, for example, come from already established companies in related fields, including pharmaceuticals, chemicals, food processing, breweries, and textiles.[14]

Established companies in declining sectors feel they must find new and more promising product lines or go out of business. Since they have a special commitment to their employees, they feel compelled to search for new areas of business. The effect of career-long employment thus differs by industry. For companies in declining sectors, lifetime employment serves as a motivation to diversify, if at all possible, into related fields of business. For companies in high-growth sectors, the effect of lifetime employment is to curb the temptation to expand into unrelated fields.

On the other hand, the pull to move into closely related areas of new product markets can be very strong. The technological opportunities and commercial incentives for product diversification tend to be greater for high technology than for old-line manufacturing. Indeed, high technology is almost by definition characterized by a high degree of technological change and commercial potential for a broad range of new product applications. It is natural, therefore, for high-technology companies to be more diversified in terms of their product portfolios.

In the information industry, for example, nearly all of the six or seven largest electronics companies in Japan are highly diversified and vertically integrated. The products manufactured by these electronic giants can be grouped into the following categories: microelectronic components (like random access memory chips and microprocessors); computers (from supercomputers to personal computers); software (from systems integration to operating software); office automation equipment (copiers, word processors); telecommunications equipment (switching devices, fiber optics and other transmission lines, value-added networks); heavy industrial applications (heavy-duty cables, wires); household appliances (refrigerators, washing machines); consumer electronics (VCRs, color televisions, hand-held calculators); and military equipment (avionics, guidance and control systems, testing equipment). The categories are closely related and constitute a natural family of products in terms of technology, marketing, and after-sales servicing. There is a logic holding together the family of diversified product markets.

The diversified nature of electronics production has created structural conditions that complicate the processes of industrial policymaking. MITI cannot coordinate the details of industrial development as readily as it could for, say, steel or the power utilities. Japan's electronics industry is too diversified and changing too rapidly to be amenable to extensive government coordination or guidance—to say nothing of central planning. Industrial policy measures that worked for old-line industries—such as the encouragement of an intra-industry division of labor—are ill suited to high-tech industries and for structural reasons simply cannot be implemented. In the early 1970's, MITI tried—and failed—to reorganize the computer industry along the lines of what seemed to be a rational division of labor. As pointed out in Chapters 1 and 2, MITI officials realize that the best industrial policy for the information industries is to focus on market-conforming measures aimed at building general consensus and to use supply-side incentives to foster competition and efficiency. A limited role for MITI also suits private industry, which does not want MITI meddling too much in areas best left to the market.

To adjust to the more diversified nature of the electronics industry, the interface between MITI and industry has had to be stretched to accommodate the need for more complex channels of communications, linking genkyoku MITI divisions with various divisions within the large, vertically integrated, diversified kaisha. However, owing to relatively high levels of market concentration among a small handful of blue-chip corporations, diversification has not proved an unmanageable problem. Despite the extended, tree-like structure of product markets, there is still an identifiable electronics industry in Japan with a fairly well defined core structure. MITI can deal with this industry roughly as it does with others—by lavishing most of its time and attention on the five or six largest companies at the top of the industrial hierarchy. The top five or six corporations contain the opinion-leaders capable of galvanizing an industrywide consensus.

Japan's "Dual-Structure" Economy

The disposition to diversify by creating independent subsidiaries or by extending long-term ties with subcontractors has reinforced Japan's so-called dual-structure economy, which consists of a small number of large parent corporations connected to a honeycomb of small and medium-sized subcontractors and subsidiaries. The ties between parent companies and independent subsidiaries or subcontractors are usually close, involving equity ownership, financing, movement of personnel from parent company to subsidiary, transfer of technology, and of course business transactions. Subcontracting is widely practiced in the field of electrical machinery, equipment, and supplies; indeed, small and medium-sized firms in this industry have one of the highest subcontracting rates of any sector in the country. In 1976, 85 percent of small and medium-sized electronics companies worked as subcontractors for big corporations, placing this field above general machinery, steel, and chemicals, and in the same range as transportation vehicles and precision equipment. Only textiles had a higher proportion of subcontractors, at 92.5 percent.[15]

From the perspective of large corporations, an extended network of closely affiliated subcontractors offers a variety of advantages, including cost reductions, semicustomized production, and flexibility to ride through business cycles. To survive, small subcontractors must be able to turn out high-quality parts at low cost, drawing on the learning-by-doing advantages of manufacturing experience and cumulative technological skills. A number of subcontractors are capable of altering product specifications and producing semicustomized parts in small quantities to meet the special needs of parent companies. Ken'ichi Imai points out that the short turnaround time needed to alter specifications "may help to explain why the Japanese manufacturing sector as a whole is capable of making quick adaptations to the changing, diversified needs of consumers."[16]

Studies have pointed out that the extensive structure of subcontracting serves as a shock absorber for the top tier of Japanese corporations in that "it allows the parent enterprise to reduce its labor costs and . . . to shift part of the risk of demand fluctuations and keep its own permanent labor force busy."[17] This buffer effect is reflected in Japan's extraordinarily high rate of bankruptcies, the vast majority of which hit the stratum of small and medium-sized firms.[18] In order for the edifice of Japan's industrial economy to hold up under the tremors of cyclical recessions, the rigidities built into the organizational structure of big corporations by heavy debt financing and lifetime employment have to be balanced by some give in the support structure of small and medium-sized enterprise. Japan's dual-structure economy is thus usually portrayed as an exploitative two-class system of production in which the second-class stratum serves the interests and needs of the large parent corporations and absorbs the shocks of business cycles.[19]

Although this characterization contains more than a grain of truth, it is too one-sided. In Japanese society, where the Confucian concept of reciprocity is deeply rooted, industrial relationships are seldom so asymmetrical in terms of either costs or benefits. Although big business certainly benefits from the structure of subcontracting, it also bears some of the costs and risks of maintaining the system. Masahiko Aoki has cast the notion of unilateral exploitation in a more complex and reciprocal light. He points out that small subcontractors and subsidiaries benefit by having access to easier credit from parent firms, which allows them to withstand business slumps.[20] Although parent firms do all they can to shield their permanent labor forces from layoffs, they are prepared to accept, if necessary, what may be very painful internal costs in terms of cutting back their temporary labor force, holding down wages, narrowing seniority-based salary differentials, and asking employees to work fewer hours in order to maintain their loyal network of manufacturing subcontractors. They do not simply pass all adjustment costs on to their small subcontractors. Precisely because they benefit from the network—to say nothing of their equity, financial, and personnel stakes in the survival of subcontractors—large parent companies are usually willing to bear their share of the cost of cyclical downturns. For risk-averse subcontractors and subsidiaries, the parent company's willingness to stabilize the ups and downs of business cycles is an asset of immeasurable value, the kind of organizational "safety net" not commonly available outside Japan.

Of course, this safety net is not always enough, given the vulnerabilities associated with smallness. The Japan Chamber of Commerce, one of the country's biggest and best-known business federations, offers a variety of support services for small and medium-sized enterprises, including financial assistance. Furthermore, the Agency for Small and Medium-Sized Enterprise within MITI provides another safety net to bolster the staying power and efficiency of small firms.

MITI's annual expenditure for small and medium-sized enterprise is the largest line item in its budget.

Looked at from an aggregate perspective, Japan's large network of small and medium-sized companies seems to create favorable conditions under which industrial policy can be administered. It is a mechanism by which large corporations stick closely to their core businesses. More important, the network contributes to Japan's industrial dynamism by lowering the barriers to new entry, diffusing technology, increasing production flexibility, and easing the headaches of costly inventory management. Japan's much-maligned dual-structure economy thus enhances the international competitiveness of the big blue-chip corporations and facilitates the administration of Japanese industrial policy.

On balance, the subcontracting system probably bestows the biggest benefits on top-tier companies. As pointed out above, however, owing to the reciprocal nature of vertical ties, it serves the interests of small and medium-sized companies as well. Among other benefits, it affords a measure of protection against the vagaries of unfettered market forces. The strength of vertical ties of interdependence reduces the dangers of costly "exit" by interlinked Japanese companies and reinforces such values as loyalty and obligated reciprocity.[21] Loyalty and reciprocity, in turn, help to convert the calculus of short-term, zero-sum trade-offs into long-term, variable-sum cooperation and collective gain.

Here is an example, in short, of how the Japanese have spun organizational safety nets to reduce the risks and costs of pure market forces.[22] Japan's extensive and complex blending of organization and market, arising from the imperfections of pure market and organizational forces,[23] offers perhaps the most compelling explanation of why industrial policy has worked. It also sheds light on Oliver Williamson's distinction between markets and hierarchies, since there is virtually no market economy where the two do not coexist in some measure.[24] The question is one of degree and of the particular organizational elements that have been built into market forces.

Keiretsu Groupings

Nowhere is the imposition of organizational structures on the market more apparent than in the well-known institution of keiretsu, Japanese industrial groupings that bring companies together in loose affiliation based on either prewar conglomerates (*zaibatsu*), financial

TABLE 3.1

Largest NEC Shareholders, March 1980

Shareholder	NEC share
Sumitomo Life Insurance	8.6%
Sumitomo Bank	6.2
Sumitomo Marine and Fire Insurance	3.5
Daiichi Life Insurance	3.4
Nippon Life Insurance	3.3
Sumitomo Electric	3.1
Sumitomo Trust and Banking	2.8
Sumitomo Trading Company	2.8
Mitsubishi Trust and Banking	2.5
Daiwa Securities	2.3
TOTAL	38.5%

SOURCE: M. Therese Flaherty and Hiroyuki Itami, "Finance," in Daniel I. Okimoto et al., eds., *Competitive Edge* (Stanford, Calif.: Stanford University Press, 1984), p. 147.

ties (*kin'yū keiretsu*), or vertical integration (*kigyō keiretsu*). Most of the leading electronics firms in Japan, such as Nippon Electric Corporation (NEC), Fujitsu, Hitachi, and Toshiba, belong to a major keiretsu. For keiretsu firms, membership actually means extensive intra-keiretsu stockholding, reliance on the main keiretsu bank for external indirect financing, and stable but by no means exclusive business transactions.[25]

Nippon Electric Corporation is a case in point. As shown in Figure 3.4, NEC belongs to the Sumitomo keiretsu. More than 30 percent of its stock is held by Sumitomo companies; Sumitomo Life Insurance and Sumitomo Bank together own nearly 15 percent (see Table 3.1). Although Sumitomo-group companies hold the largest share, financial corporations from other keiretsu groups, such as Mitsubishi, also own equity shares. This brings to light a noteworthy point: keiretsu groups are not closed, monolithic superstructures tightly held together by a single holding company, as were prewar zaibatsu.[26] They do not operate as fully self-contained or mutually exclusive entities. Significant cross-keiretsu ties in terms of business transactions and equity shareholdings prevent the industrial groupings from degenerating into isolated blocs, a development that would rigidify Japan's industrial structure and rob it of much of its efficiency, flexibility, and dynamism. What exists instead is a loosely knit, permeable set of industrial networks, connected through cross-cutting linkages—what Aoki calls "quasi-tree structures."[27]

This is not to say that keiretsu structures lack cohesion and mean-

The Sumitomo Group

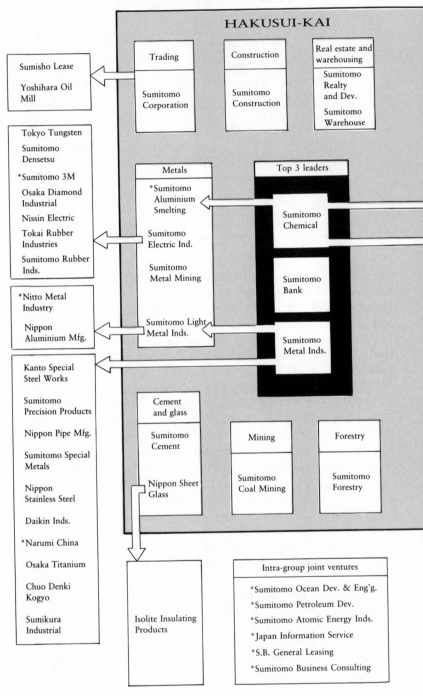

HAKUSUI-KAI

Sumisho Lease

Yoshihara Oil Mill

Tokyo Tungsten

Sumitomo Densetsu

*Sumitomo 3M

Osaka Diamond Industrial

Nissin Electric

Tokai Rubber Industries

Sumitomo Rubber Inds.

*Nitto Metal Industry

Nippon Aluminium Mfg.

Kanto Special Steel Works

Sumitomo Precision Products

Nippon Pipe Mfg.

Sumitomo Special Metals

Nippon Stainless Steel

Daikin Inds.

*Narumi China

Osaka Titanium

Chuo Denki Kogyo

Sumikura Industrial

Trading

Sumitomo Corporation

Construction

Sumitomo Construction

Real estate and warehousing

Sumitomo Realty and Dev.

Sumitomo Warehouse

Metals

*Sumitomo Aluminium Smelting

Sumitomo Electric Ind.

Sumitomo Metal Mining

Sumitomo Light Metal Inds.

Top 3 leaders

Sumitomo Chemical

Sumitomo Bank

Sumitomo Metal Inds.

Cement and glass

Sumitomo Cement

Nippon Sheet Glass

Mining

Sumitomo Coal Mining

Forestry

Sumitomo Forestry

Isolite Insulating Products

Intra-group joint ventures

*Sumitomo Ocean Dev. & Eng'g.

*Sumitomo Petroleum Dev.

*Sumitomo Atomic Energy Inds.

*Japan Information Service

*S.B. General Leasing

*Sumitomo Business Consulting

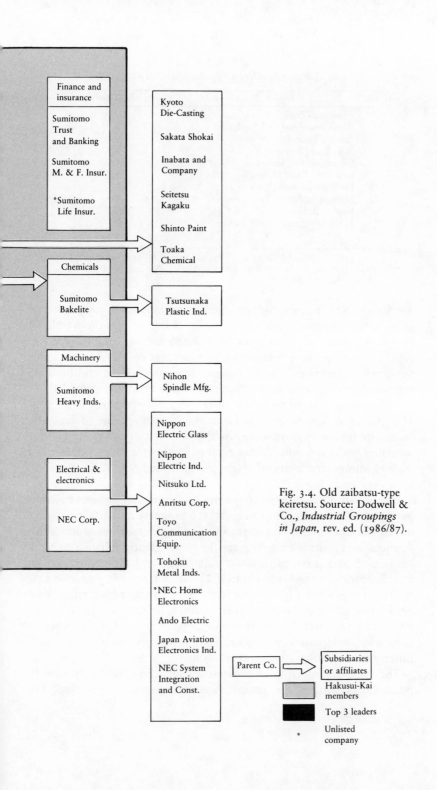

Fig. 3.4. Old zaibatsu-type keiretsu. Source: Dodwell & Co., *Industrial Groupings in Japan,* rev. ed. (1986/87).

TABLE 3.2

Largest Lenders to NEC, March 1980

Lender	NEC loans
Sumitomo Bank	16.7%
Sumitomo Trust and Banking	11.4
Japan Export-Import Bank	6.7
Yokohama Bank	4.9
Sumitomo Life Insurance	4.6
Industrial Bank of Japan	4.3
Long-term Credit Bank of Japan	4.3
Mitsubishi Bank	4.0
Kyowa Bank	3.9
Japan Bond and Credit Bank	2.3
TOTAL	63.1%

SOURCE: M. Therese Flaherty and Hiroyuki Itami, "Finance," in Daniel I. Okimoto et al., eds., *Competitive Edge* (Stanford, Calif.: Stanford University Press, 1984), p. 147.

ing. Just as NEC's stockholdings suggest that Sumitomo is its center of gravity, so, too, its pattern of bank borrowing reflects the weight of Sumitomo financial institutions. In 1980, Sumitomo Bank served as NEC's largest creditor, extending 16.7 percent of NEC's total loans outstanding, or roughly $188 million. Sumitomo Trust and Banking lent another 11.4 percent, and Sumitomo Life Insurance provided 4.6 percent, bringing the total for the three to 32.7 percent (see Table 3.2). The greater weight of bank loans in the overall scheme of Japanese corporate financing is indicated by the fact that in 1975, marketable securities (stocks, bonds, commercial papers) came to 35.9 percent of total nonmonetary financial assets in the United States, compared to 12.9 percent in Japan.[28]

U.S.-Japan differences in corporate financing can be seen by comparing the various sources of capital for Japanese and U.S. producers of semiconductors. As Table 3.3 illustrates, the two outstanding differences in corporate financing are the much greater reliance on bank loans in Japan (external indirect financing) and the greater use of internal funds in the United States. The data since 1975 suggest that the trend is toward a narrowing of the gap, a diminishing reliance on bank loans and greater use of internal funds by Japanese firms. However, even if the trend is toward some degree of national convergence, complete similarity seems very unlikely because of enduring structural differences in banking-business relations.

The combination of intra-keiretsu stockholding and heavy debt financing accentuates the importance of keiretsu membership. For

TABLE 3.3

Comparative Sources of Corporate Financing: Japan and the United States, 1970–1979

Period average	Internal funds		Equity issues		Bond issues		Loans	
	Japan	(U.S.)	Japan	(U.S.)	Japan	(U.S.)	Japan	(U.S.)
1970–74	47.2%	(61.8%)	2.9%	(19.6%)	8.8%	(9.4%)	41.1%	(9.2%)
1975–79	55.2	(79.0)	7.3	(6.6)	10.7	(8.6)	26.8	(5.8)
1970–79	51.2	(70.5)	5.1	(13.0)	9.7	(9.0)	34.0	(7.5)

SOURCE: Same as Table 3.1, pp. 142–43.

NEC, the benefits of being part of the Sumitomo group are substantial. Corporate stockholders like Sumitomo Bank are willing to accept low rates of return on investment (ROI) as long as the real value (as distinct from par value) of equity shares appreciates. NEC is thus under less compulsion than most U.S. firms to adopt strategies yielding high short-term profits, as pointed out in Chapter 1. If U.S. firms fail to earn regular profits, the value of their stock plummets, leaving them vulnerable to the threat of outside takeover. Interlocking stockholdings by corporations willing to accept low ROI give Japanese companies some degree of insulation from the imperatives of equity markets.

Japan's pattern of intra-keiretsu stockholding and debt financing also gives NEC some measure of protection against downside risks. Because Sumitomo Bank is both a shareholder and a lender, its relationship with NEC is much closer than that between U.S. commercial banks and companies. It is a long-term relationship of interdependence characterized by mutual trust and obligation, not the legalistic, arms-distance relationship characteristic of banking-industry transactions in the United States.[29]

The relationship helps corporations cope with the structural consequences of Japan's past deficiencies in capital markets, especially the vulnerabilities associated with high debt financing. In hard times, Sumitomo Bank is often willing to refinance old loans or extend new ones; in return, NEC is expected to give full disclosure of information concerning company operations, cooperate when business is good, and, if necessary, accept some infringement of autonomy in allowing the bank the right to send its own people into key management positions during a crisis. As in the case of parent company–subcontractor ties, the structure of banking-business relations accentuates the benefits of cooperation, loyalty, long-term commitment, reciprocity, and "voice" by closing off easy points of "exit" and superimposing an orderly organizational framework on the interplay of market forces.[30]

The system of risk diffusion and mutual assistance also extends to buyer-seller relationships of long standing, both within and across keiretsu networks. Japanese companies often operate on the basis of an implicit hierarchy of customers. The oldest and biggest are at the top in terms of financing, delivery schedules, customer service, and first claim on supplies when shortfalls occur (as in the case of semiconductor memory chips). Buyers and sellers conduct business within the framework of implicit long-term contracts, not on the basis of spot transactions in a completely open-ended marketplace.[31] Very often, buyers and sellers hold shares in one another's companies, a structural link that has the effect of consolidating common interests and a commitment to long-term relations.[32]

Thanks to the existence of such organizational ties, marginal adjustments in short-term transactions with respect to product quality, quantities, price, financing, and service can be made. When times are bad for buyers, sellers can adjust the terms of transaction; since a running balance sheet of favors and obligations is maintained, sellers can expect to be repaid with similar treatment at some point in the future. Accordingly, the mutually accommodating nature of the buyer-seller relationship can help to stabilize business transactions. As end users, Sumitomo-related companies, for example, may have strong incentives to purchase NEC products rather than Mitsubishi or Mitsui products; on the other hand, since transactions across keiretsu boundaries are extensive, the force of extramarket factors should not be exaggerated, especially in high technology.

The keiretsu framework thus provides an effective mechanism for diffusing and diminishing risks in Japan's industrial system. Indeed Iwao Nakatani argues that the "primary purpose of [keiretsu] group formation seems to be the sharing of risks and profits among group members by which stabilization of corporate performance is enhanced."[33] Comparative data reveal that the performance of keiretsu firms tends to fall below that of independent companies on such standard indicators as profitability and stockholder dividends. That keiretsu firms earn lower profits and pass on smaller dividends to shareholders seems puzzling, since they tend to be blue-chip companies with stable sales networks and control over product prices through market dominance. Given these circumstances, neoclassical economics would predict higher profits and dividends. The reason they are not higher,

Nakatani hypothesizes, is that the firms maximize the joint utility of their corporate constituents—employees, lending banks, stockholders, subsidiaries, and others.[34] In exchange for the insurance coverage, keiretsu companies are willing to give up some portion of their profits. The marginal loss is a small price to pay for the leveling off of business fluctuations and for the comprehensive coverage provided by mutual insurance schemes against downside risks. The effectiveness of such joint insurance schemes is seen in the relatively small number of bankruptcies of keiretsu-anchored big corporations compared to those of independent companies.

The effectiveness of keiretsu-based insurance schemes in Japan is also reflected in a comparison of the Chrysler and Mazda bailouts: the former required U.S. government intervention, whereas the latter was handled by the Sumitomo keiretsu group, especially Sumitomo Bank, and Tōyō Kōgyō's (Mazda's) subcontractors and distributors.[35] If the Chrysler-Mazda comparison is typical, the Japanese industrial system of risk diffusion and crisis management, organized around keiretsu structures, intercorporate stockholding, and bank-industry interdependence—all of which encourage long-term commitment, loyalty, and "voice" over fast market exit—seems to provide an effective web of safety nets within the private sector. The existence of multilayered safety nets in the private sector is a boon to the Japanese government, since it relieves MITI of the onerous task of bailing out faltering companies. To sustain a lean and forward-looking industrial policy, MITI cannot allow itself to be dragged into the costly and politicized business of corporate bailouts, as the governments of France and England have been. In this sense, keiretsu networks have lightened the policy burdens that would otherwise weigh heavily on MITI.

Keiretsu structures also help to impose organizational order on market competition. Surprisingly, Japan's organizational mechanisms for cooperation have not dampened the gusto with which companies compete in the marketplace. Indeed, by helping to free corporations from the short-term dictates of equity markets, the structure of keiretsu financing and intercorporate stockholding may actually serve to intensify market competition; at the same time, however, it supplies some incentives to cooperate and the organizational framework within which cooperation can take place.

By bringing top executives from member companies together for regular meetings and by opening a variety of communication channels, keiretsu groups also provide a network for interfirm communication.

From MITI's point of view, the existence of this communication network is useful not only for transmitting and receiving information but also for mobilizing consensus on specific aspects of industrial policy. For large projects requiring multifirm participation, such as the design and construction of large manufacturing facilities, loose keiretsu networks and sometimes cross-keiretsu alliances can be used to organize suitable teams, especially through the lead of keiretsu banks.

On the other hand, there are definite costs associated with the existence of keiretsu networks, and these must be weighed against the advantages. Perhaps the most troubling cost is related to the antitrust implications of keiretsu groupings. When giant producers belong to a small number of industrial groups representing the lion's share of Japan's manufacturing output, when the level of market concentration for specific industries is high, and when there are interlocking patterns of corporate stockholding, the temptation to collude tends to be very strong. Opportunities for collusion are almost built into the structure. Richard Caves and Masu Uekusa point out that in the past, Japanese corporations frequently engaged in such collusive practices as price fixing, particularly in old-line industries where minimum prices had been established during cyclical downturns.[36] For highly leveraged firms in capital-intensive industries, the impulse to collude is particularly strong, and oligopolistic behavior is facilitated by the fact that leading firms own stock not only in related companies but sometimes in competitor firms as well.

The interlocking pattern of stockholding also tends to exacerbate international trade frictions by making it difficult for foreign firms to acquire Japanese companies and break into Japanese home markets.[37] The logic of pure market principles, such as lower prices, may not be enough to win foreign firms a foothold in Japanese markets, since marginal price differentials may be more than offset by organizational imperatives, including long-standing business relations. A preference for dealing with fellow Japanese companies in which one holds equity shares and with which one has had long-standing business dealings may make perfectly good sense from a transaction point of view, but the logic of organizational linkages introduces formidable barriers to outside entry for non-Japanese producers.

In addition to the inefficiencies caused by occasional acts of collusion, the existence of keiretsu groupings can also lead to such undesirable outcomes as the installation of excess plant capacity; this is the consequence of a desire on the part of each keiretsu to have at least one member company represented in each major industry, a situation referred to as "one-set-ism." Excess capacity, in turn, can give rise to excessive competition; as pointed out in Chapter 1, this generates pressures to overproduce during recessions and to export surplus production in order to climb out of recessions.

Keiretsu networks may be changing as big corporations respond to the changing requirements of high technology by shifting more of their R&D resources to small subsidiaries. Since 1974, the trend toward R&D decentralization has shown a marked increase, with anywhere from one-quarter to one-half of R&D activities being transferred from large parent corporations to small subsidiaries.[38] The dispersal of money and manpower to quasi-independent subsidiaries is, from the parent companies' perspective, a means of enhancing R&D flexibility while retaining for themselves the advantages of lower production costs and scale economies. Whether the small subsidiaries will be able to convert this windfall of R&D resources into significant advances in technology remains to be seen.

The proliferation of subsidiaries and the shifting of R&D effort from large parent corporations to smaller subsidiaries indicates that keiretsu organizations are passing through a period of transition. The transformation is taking place largely under the impact of two developments: the deregulation of Japan's financial system and the rapid pace of technological change introduced by the shift to high technology. Old keiretsu boundaries, never very clear to begin with, are being blurred further by the force of such technological and commercial developments as the cross-licensing of technology, second-sourcing and original equipment manufacturing (OEM) agreements, and joint ventures. Subcontractors are no longer tied into exclusive arrangements with parent firms. The trend is toward a diversification of sales. The emerging structure of subcontracting and inter-keiretsu relationships is thus complex, with crisscrossing ties of interdependence blurring organizational boundaries between industrial groupings.

Signs of change are also observable inside keiretsu groups. Given the latitude for horizontal diversification in high technology and for new entry up and down the ladder of vertical integration, the idea of having one complete set of companies within each major keiretsu is no longer as attractive as it used to be. Member companies are starting to enter some of the same product markets. Of course, intra-keiretsu competition can be mitigated, to some extent, by a division of labor of the kind observable between NEC and Sumitomo Electric, two sister members of the Sumitomo keiretsu specializing in different areas of electronics. But not all internal competition can be so readily contained. Compelling incentives for sister companies to diversify into overlapping fields may be hard to resist, and this could further erode the internal cohesiveness of keiretsu organizations.

The momentum for financial deregulation, which is coming from two sources—the ballooning secondary market for deficit-covering government bonds and foreign pressures to lift regulatory financial controls—is also altering the role of Japanese banks and the nature of banking-industry relations. As regulatory controls are gradually lifted, Japanese corporations are beginning to reduce their high levels of dependence on bank lending by relying more heavily on retained earnings and raising more equity on foreign and domestic capital markets. To what extent these trends diminish the power of Japanese banks bears close watching, since the banking-industry relationship has been the linchpin of Japan's industrial structure. Should that link weaken, a sequence of changes could be set off.

Japanese Financial Markets

Regulated Financial System

In his book *Governments, Markets, and Growth*, John Zysman seeks to establish a causal link between the nature of a country's financial system and the state's capacity to intervene effectively in the economy. Specifically, he argues that the scope, nature, and capacity of an industrial state to intervene depends on the structure of its system of capital allocation.[39] In countries with highly developed capital markets, like the United States and Great Britain, the state is constrained in what it can do by the limited range of financial tools at its disposal. The arm's-length nature of corporate financing through impersonal capital markets forces companies, banks, and government into separate spheres of activity, with clear and distinct boundaries keeping the state from crossing over any time it wishes. Industrial progress in mature capital-market systems is spearheaded by private enterprise, with the state capable of playing a supporting but only secondary role.

In countries where capital is channeled largely through banks, like Japan and France, the state is in a position to extend a highly visible and vigorous hand in the operation of the economy. The system of external indirect financing gives the state the power to ration credit to banks and channel capital to targeted sectors. Not only does this enable the state to influence the directions of industrial change; the very fact that it possesses such power almost compels the state to play an interventionist role. Since capital is the lifeblood of industrial activity, the state's capacity to set the supply and channel the flow of capital constitutes one of the most effective tools of industrial policy. Zysman argues that in credit-based systems, therefore, the state, not private companies, assumes the initiative for industrial change.

In credit-based systems, private companies lack the autonomy to keep the state out of their spheres of activity. The lines of demarcation between the private and public sectors are hard to maintain. Because of their heavy reliance on bank borrowing, Japanese and French companies are structurally more dependent on banks and on the state than their counterparts in market-based financial systems.[40] In Japan, the origins of this structural dependence can be traced to the tight regulatory controls imposed on the country's financial system during the era of wartime mobilization.[41] Japanese industries came to depend on the state not only for preferential loans and R&D subsidies but also for countercyclical macroeconomic measures and predictable, low interest rates. For companies with high debt-to-equity ratios, even marginal fluctuations in interest rates can have an enormous impact on debt-servicing burdens. Hence, a country's financial system seems to affect both the need for and the state's capacity to undertake market intervention. It also structures, as Zysman points out, the nature of the relationship between the state and private enterprise.

Although it is true that Japan's credit-based financial system has bestowed formidable powers on the state vis-à-vis the private sector, these powers have waned progressively as Japan's capital markets have developed and as its financial system has become internationalized. The Japanese government's share of industrial financing fell from a high of 28.3 percent in the early 1950's to roughly 13 percent by the mid-1970's, with only 3.3 percent of total state financing earmarked for industrial targeting. Moreover, with private savings exceeding business investment demand, Japan is no longer strapped by the scarcity of capital that spawned the old system of credit rationing. Nor do priority industries rely on the announcement effects of MITI's industrial policy to secure the full bank backing they need to meet their ambitious schedules of capital investments. Data on corporate financing reveal a noteworthy trend away from debt financing toward heavier reliance on equity financing. Clearly, the once-prominent role of the state in industrial financing has diminished very rapidly as Japan's financial system has changed. It is likely to continue to wane as the trend toward financial deregulation moves ahead.[42]

Just as the state's financial role has shrunk, so too has the scope of its authority. The power to ration credit has devolved, to a large extent, on a multitude of actors dispersed within the private sector. This devolution is symptomatic of a much broader shift in the balance of power between the state and private enterprise, one in which the relative autonomy of the private sector has increased with the growth of the economy. Following three decades of pace-setting expansion, Japan's private sector is bigger and stronger than ever. Key industries have achieved world preeminence; a variety of new industries have emerged; and interest groups have mushroomed everywhere—while the once-imperial power of the state has atrophied.

The "Relational State"

This is not to imply that MITI is no longer powerful. Quite the contrary; on any comparative basis, it remains perhaps the single most effective institution for industrial policy-making among all the major states in the Organization for Economic Cooperation and Development (OECD). However, the extraordinary power it once exercised during the 1950's and 1960's is definitely a thing of the past, owing to the maturation of Japan's economy. MITI's influence today rests less on control of credit allocation or the possession of formal authority than on its powers of persuasion and coordination vis-à-vis the private sector, its capacity to gather and process information, set directions and priorities, promote private-sector and collective interests, and serve as an intermediary between domestic and international economies.

MITI and industry see themselves bound inextricably together in a long-term relationship. Government-business relations are not conducted at arm's length, with free entry and exit, as in the United States. In Albert Hirschman's terms, any "exit" from the long-term commitment would be very costly, for it would destroy the foundation of mutual trust on which the commitment is based.[43] In consequence, both sides possess a voice in the formulation of industrial policy, in keeping with the spirit of consensual policy-making; both feel bound by a sense of common interest.

This suggests that the Japanese state should not be conceived of simply as an administrative apparatus in charge of regulating private industry. Its power cannot be fully fathomed in terms of such standard indicators as its autonomy from interest-group pressures, capacity to change the behavior of private-sector groups, or ability to change the structure of the society within which it operates.[44] Nor, conversely, can the private sector's power be measured by its capacity to resist the state's attempts to impose its will on the rest of society. Casting government-business relations in Manichaean terms and raising the question of which of the two is stronger misses the subtlety and complexity of the relationship. Although the Japanese state is strong, as various observers have pointed out,[45] so too is the private sector. Their relationship is not adversarial or a tug-of-war. They try to pull together in the direction of achieving common goals.

Without industry's willing cooperation, the Japanese state would not be nearly as powerful or effective. Instead of labeling Japan a "strong" state, therefore, perhaps it would be more accurate to call it a "societal," "relational," or "network" state, one whose strength is derived from the convergence of public and private interests and the extensive network of ties binding the two sectors together. Even at the zenith of its power, MITI always worked in close concert with private industry. It rarely had to confront private industry in a showdown of strength, thanks to the thoroughness of prior consultations and the willingness of both sides to give a little. Now that MITI has lost a number of its most extraordinary sources of powers—such as foreign exchange allocations, foreign direct investments, and technology licensing—close working relations with private industry have become more important than ever. To elicit cooperation from private enterprise, MITI draws on a combination of resources: superior information, economic logic, long-term vision, the capacity to mediate and coordinate, promotion of producer interests, and its mandate to safeguard collective and national interests.

Market Concentration

MITI's efforts to work with the private sector are facilitated by the comparatively high degree of market concentration across a range of industries, especially the basic, heavy industries. Looking at market concentration without reference to keiretsu membership, Caves and Uekusa conclude that the average concentration ratio for manufacturing sectors in Japan is not very different from that in the United States. The "overall indicators show no strongly sustained long-term trend in concentration."[46] However, when companies are grouped into keiretsu networks, the Japanese concentration ratios rise dramatically, far beyond U.S. levels. Table 3.4 sets forth the keiretsu-adjusted data, covering the country's six largest keiretsu groups. As indicated in the table, the six leading keiretsu groups hold dominant market shares in

TABLE 3.4
Keiretsu-Adjusted Market Shares, 1974

Sector	Mitsui	Mitsu-bishi	Sumi-tomo	Fuyo	Sanwa	Daiichi	Total
Rubber	0%	4.9%	45.4%	13.4%	17.1%	18.6%	100%
Shipbuilding	8.7	38.3	0	1.3	10.0	39.0	97.3
Electrical machinery	19.0	10.8	38.3	19.7	1.0	8.1	96.9
Nonferrous metals	43.4	20.0	16.3	3.6	0	9.8	93.1
Textiles and apparel	26.3	18.4	13.8	8.2	19.7	0.3	86.7
Drugs	11.4	1.9	52.5	1.4	15.0	4.1	86.3
Petroleum	6.2	13.4	0	40.1	15.9	8.3	83.9
Construction	11.0	5.8	21.3	17.2	12.5	14.2	82.0
Chemicals	17.6	27.6	13.1	10.3	11.0	2.2	81.8
Machinery	10.5	10.4	33.8	7.4	8.9	8.3	79.3
Wire and cable	9.4	9.0	24.5	0	0	26.0	68.9
Precision equipment	0	36.2	5.1	10.7	0	11.5	63.5
Metal products	25.0	8.3	11.0	2.1	2.9	13.4	62.7
Steel	2.1	0.8	13.1	18.6	14.6	11.1	60.3
Motor vehicles	31.3	8.0	9.7	3.1	2.3	0.3	54.7

SOURCE: Ken'ichi Imai, "Interfirm Group Behaviors in Japanese Industrial Organization" (unpublished paper, Feb. 16, 1979), p. 6.
NOTE: Italics indicate industry leader.

a number of industrial sectors in Japan, especially those that convert raw materials into key intermediate goods such as rubber, nonferrous metals, petroleum, and chemicals.

Caution should be exercised in interpreting the significance of the data, however. One should not, for example, jump to the conclusion that Japan's economy is completely controlled by monopolies and oligopolies. Without the centralized holding companies that existed in prewar zaibatsu, the keiretsu groups do not function as monolithic units; the lack of cohesion is especially apparent in the case of the Sanwa, Daiichi, and Fuyo groups, where interfirm ties are very loose. Although Toshiba is listed as belonging to the Mitsui keiretsu, to cite another example, the affiliation is so loose that the categorization is almost without meaning; innumerable other examples of loose affiliation can be cited. The keiretsu-adjusted data therefore overstate the actual degree of market concentration. Moreover, competition between firms is generally so intense that oligopolisitic, inter-keiretsu collusion can be ruled out (except perhaps in cases of collective crises). And with strong companies standing outside the umbrella of the six leading groups, the industrywide coordination necessary for successful and sustained oligopolization is in most cases impossible.

On the other hand, the significance of very high keiretsu-adjusted concentration ratios should not be dismissed. They are especially significant in mature, capital-intensive industries with heavy sunk investments in plant capacity, where economies of scale are the prime determinants of commercial competitiveness. Companies in such industries undoubtedly have strong incentives to collude, especially when business is slow, and if the number of producers is small, it is possible to coordinate interfirm behavior. Moreover, MITI's disposition to regulate excessive competition has given companies in basic manufacturing industries regular opportunities to bring unruly competition under harness.[47]

From MITI's point of view, high levels of concentration have also had some beneficial effects in terms of administering industrial policy. They have expedited industrial catch-up, for example, by permitting large corporations to capitalize on economies of scale. Not only did these large companies control dominant shares of Japan's domestic markets—often over 50 percent—behind the walls of infant industry protection, they also benefited from preferential treatment made possible by Japanese industrial policy, including preferential financing and access to the country's best college graduates. Heavy market concentration in the basic manufacturing industries can thus be considered both an outgrowth of Japanese industrial policy and a means of achieving Japan's early postwar objective of industrial catch-up.

Sector-specific consensus, the sine qua non of Japanese industrial policy, is facilitated by the combination of high market concentration and the frequency with which a company leader—usually the one with the largest market share—can be identified. This combination enables an industry to iron out differences of opinion within its own ranks and to take a unified stand on issues requiring rapid implementation of industrial policy. In the process of reaching consensus, of course, MITI is usually involved as an objective arbiter, a neutral coordinator, or an advocate of certain policy measures. But unless irreconcilable conflicts prevent it from doing so, each industry will try to work out differences among individual companies on its own. Self-regulating interest aggregation within industry is certainly a boon for the Japanese government, if only because it reduces its work load.

The market shares of industry leaders are shown in italics in Table 3.4. Usually, the leading firm(s) within the dominant keiretsu holds at least a 33 percent market share. Note that no single keiretsu is dominant in all fields or even in a majority of them. Leadership is fairly well dispersed. The three oldest and largest keiretsu, Mitsui, Mitsubishi, and Sumitomo, hold leadership positions in by far the largest number of fields. Except for shipbuilding, where Daiichi is strong, and petroleum, where Fuyo is preeminent, virtually all other fields are evenly balanced or Mitsui, Mitsubishi, or Sumitomo holds a commanding market share. It looks almost as if a tacit division of labor has been worked out, an implicit agreement to maintain a rough balance of power by allowing the three to carve out spheres of influence. But there is, of course, no evidence that the three keiretsu groups ever got together to reach such an accord, much less that MITI planned it this way. The more likely explanation is that in competing vigorously across all sectors, each keiretsu came to concentrate on particular fields in which it either held or came to hold a comparative advantage.

Whatever the explanation, the fact is that markets in many sectors are highly concentrated, with an acknowledged leader or first among equals among competing companies and keiretsu groups. It is not a contradiction in terms to call this pattern a "competitive hierarchy." The existence of an acknowledged first among equals, a clear pacesetter and opinion leader, greatly facilitates the task of consensus building within each industry. By working out mutually acceptable agreements with the opinion leader, MITI can often secure full cooperation from industry as a whole. Here again is an illustration of how the structure of Japan's private sector, despite the dangers of collusion and rigidity, has simplified the task of formulating and administering industrial policy.

Concentration ratios in high-technology sectors vary by industry. They are low in such industries as machine tools and software but relatively high in industries like aircraft and semiconductors. The concentration ratios in integrated circuits (ICs), by type of device, are shown in Table 3.5. Although concentration ratios are high, the competition between large, diversified electronics companies is fierce. Moreover, the competitive hierarchies are in constant flux, as wave after wave of new products hits the market. This is not true in heavy manufacturing industries like steel and automobiles. Product life cycles in high technology are typically much shorter than in the old-line sectors. Rankings of company and keiretsu leaders by product markets are constantly changing.

The antitrust implications of market concentration ratios, as Kenichi Imai points out, vary across product markets in accordance with the opportunities for differentiation.[48] If the leeway for product dif-

TABLE 3.5

Concentration Ratios by IC Device

Device Type	Top 5 concentration ratio	Top 5 manufacturers
Linear ICs	70%	NEC, Hitachi, Matsushita, Toshiba, Sanyo
Bipolar ICs	80	Hitachi, Fujitsu, NEC, Toshiba, Texas Instruments
MOS logic ICs	75	NEC, Hitachi, Toshiba, Sharp, Mitsubishi
MOS memory ICs	90	NEC, Hitachi, Fujitsu, Toshiba, Mitsubishi

SOURCE: Nomura Research Institute, *Japanese Semiconductor Industry Report* (Tokyo, 1983), p. 9.

ferentiation is broad, as is the case for electrical machinery, the dangers of collusion are not as great as they are in product markets like rubber or steel, where the opportunities for differentiation are limited. Accordingly, the pressures to collude are lower because producers can find niches in which to specialize.

In undifferentiated markets, however, where competition is zero-sum and demand is inelastic—as in the case of most raw material–converting, intermediate goods industries—the pressures to collude by forming informal cartels or by coordinating production levels and prices tend to be very strong. It is in these industries that interfirm relations assume the form of extensive intercorporate stockholding, linking competitors more closely than membership in looser keiretsu networks. Indeed, the whole issue of intercorporate stockholding in undifferentiated product markets (such as rubber, urea, bearing steel, rails, and so forth) raises complex and disturbing questions about possible violations of antitrust.

Multiple Access Points

High levels of market concentration, reinforced by interlocking patterns of stockholding, have thus facilitated government-industry coordination, just as keiretsu networks, subcontracting structures, and financial regulation have done. Whether, in the absence of such organizational structures, MITI could have orchestrated industrial policy as deftly as it did is a question that cannot be definitively answered. However, if the arguments advanced here are valid, the answer would probably be no. The imposition of extra-market institutions like keiretsu on the marketplace has had the effect of giving the government an unusually broad range of "access points" by which to influence the directions of the private sector. The "access points" include:

1. Excellent channels of vertical communications via subsidiary and subcontracting networks
2. Ready-made horizontal forms for consensus building through keiretsu groupings
3. A small number of big corporations with commanding market shares, often including a clear leader willing to take the initiative in forging intra-industry consensus
4. Clearly identifiable liaisons within private corporations that stay well focused on core businesses
5. Industrial associations for intra-industry interest aggregation
6. Opportunities for interfirm communication and mobilization through intercorporate stockholding
7. The availability of a huge source of funds in the Fiscal Investment and Loan Program (FILP), which is well insulated from the pressures of parochial interest-group lobbying and can be used for social overhead investment projects and industrial targeting
8. Government financial institutions, like the Japan Development Bank, capable of allocating capital to priority sectors and, to some extent, of reducing the perceived risks of extending commercial loans to these sectors
9. Until 1985, a heavily regulated financial system with underdeveloped capital markets in which the Ministry of Finance and the Bank of Japan have wielded wide-ranging powers, including the determination of interest rates
10. Highly leveraged corporations vulnerable to perturbations in the business environment, especially cyclical downturns, and therefore highly dependent on banks and favorable government policies
11. A dozen powerful city banks with very close ties to and considerable influence on the behavior of private corporations
12. A half-dozen giant trading companies in control of well over half of Japan's world trade, medium-term financing for commodity transactions, and distribution networks
13. Over a hundred public corporations, a number of which are strategically placed and influential

The Japanese government can thus choose to intervene through a wide range and variety of access points. This allows for selective and close intervention, finely tuned to meet the needs of specific industries. Most of the levers identified above are outgrowths of Japan's distinctive blend of organizations and markets. They give the government the flexibility to tailor policies to the needs of each sector.

To put matters in comparative perspective, it should be mentioned that several access points can be found in other political economies, such as the cluster of public corporations in the United States and highly leveraged corporations in France. Moreover, the Japanese government lacks a few institutional levers that exist prominently elsewhere, such as huge procurement budgets (the United States) and nationalized companies (France, Italy, and Norway). But the dangers of politicization and economic inefficiency associated with huge procurement budgets and nationalized companies are sufficiently great that Japanese government officials do not bemoan their absence. By contrast, the structural levers that MITI uses to intervene—such as vertical communication networks—entail substantially less danger of politicization or semipermanent government involvement.

Understanding that the Japanese government has the option of stepping in when circumstances call for it and that MITI can choose one or more convenient access points of entry clears up the conundrum mentioned in Chapter 1: namely, how the Japanese government can be both a minimalist state with a hands-off attitude toward the market economy and at the same time an interventionist state that extends its long arms to steer the market when it veers off course. The range of access points permits the Japanese government to stand back and watch the market function while retaining the ability to intervene at any time. It also stretches the range of industrial policy instruments available.

The potential for timely state intervention is somewhat analogous to the relationship between Japanese banks and industry. Under normal circumstances, Japanese banks adhere to a hands-off policy, providing advice and support for private companies but letting them operate independently. However, when companies fall into serious trouble, leading banks step in and take an active part in managing the companies until they are out of trouble. Obviously, the MITI-bank analogy—suggesting structural parallelism or isomorphism at two levels— should not be taken too far, but the leeway to intervene pragmatically as circumstances warrant is similar. In both cases, the points of access are by-products of extra-market organizational structures superimposed on the free-exit open-endedness of the market.

Given the range of strategic access points, it is remarkable that the Japanese state has exercised as much self-restraint as it has in the face of what must be strong institutional impulses to intervene often, widely, and for indefinite periods of time. The self-restraint can be ascribed to several factors: (1) vigorous market competition, especially in high technology, which MITI is careful not to stifle; (2) effective organizational mechanisms for risk diffusion within the private sector, which obviate much of the need for state intervention; (3) the private sector's capacity, therefore, to function at a high level of efficiency without government meddling; and (4) the perceived high costs associated with unnecessary intervention.

The private sector's capacity for self-regulation is derived not simply from the work of Adam Smith's invisible hand but, more important, from the channeling of market forces through an institutional framework that structures their interplay and moderates the harsh and sometimes unacceptable outcomes of decentralized, "pure" market competition. It is the same intersection of market and organization that provides access points for state intervention, which MITI wisely chooses to utilize in moderation. One reason that the Japanese government possesses so diverse a repertoire of industrial policy instruments is precisely that the structure of access points makes sector- and situation-specific intervention feasible.

Intermediate Organizations: Policy Networks

We have analyzed MITI and the structure of private enterprise, finding features that help explain Japan's remarkable capacity for government-business coordination. However, neither the effectiveness of Japanese industrial policy nor the unusual closeness of government-business relations can be fully understood by reference to MITI and private industry alone. What is needed to complete the picture is an exploration of the maze of connections that link public and private sectors together in what is called here the "intermediate zone." This is the network of ad hoc, informal ties—not just MITI's formal organization or the structural characteristics of Japanese industrial organization—that gives industrial policy and government-business interactions the resilience and adaptability for which Japan is renowned.

Peter J. Katzenstein refers to the formal and informal networks that bring government officials and leaders from the private sector together to formulate public policy as "policy networks."[49] The density of these networks is a function of the political structures in which they are embedded; specifically, the number and importance of policy networks vary with the level of government and business centralization and the degree of differentiation between state and society. The greater the centralization and the lower the degree of differentiation between state and society, the denser the thicket of policy networks, or at least the greater the leeway for their establishment.

Taking Japan as an example, we can observe that the combination of centralization and state-society interpenetration has created conditions more conducive to effective state intervention than those of most other industrial states. The leeway to build, reconstruct, or phase out informal policy networks—networks arising not simply from short-term self-interest but out of the framework of long-term, obligatory, and affective ties—seems to give Japan's political economy extraordinary flexibility.

Except for the work of scholars like Chalmers Johnson and Ken'ichi Imai,[50] the literature on Japanese government-business relations has paid only scant attention to the role of these intermediate organizational networks. The omission is hard to understand, given the structural and functional importance of public-private linkages. It is largely owing to the existence of such networks that government officials and business leaders are able to transcend the narrow confines of membership in formal institutions.

Policy networks can be divided into two types, formal and informal. The formal type consists of various public corporations (such as NTT and Japan National Railway), public enterprises (such as Japan Housing Corporation), and an assortment of other nonprofit organizations (such as the Information Technology Promotion Agency and the JECC).[51] Imai divides intermediate organizations into two groups: quasi-governmental organizations (public corporations and public enterprises), like NTT, which operate under strict government regulation, and quasi-nongovernmental organizations (other nonprofit organizations), like JECC, which operate more like private entities, relatively independent of government directives.[52] In Japan, there are 112 quasi-governmental organizations, if we use a narrow definition of public corporations and public enterprises, and several thousand quasi-nongovernmental organizations.

These intermediate organizations perform an absolutely indispensable role in the Japanese economy. They function as regulated natural monopolies, raising revenues, making developmental investments, building social overhead infrastructure, channeling funds for industrial targeting, redistributing income to the less productive sectors, gathering information from abroad, promoting trade and investment, carrying out industrial research, organizing national projects, planning regional development, advancing particular industries and technologies, regulating the price of certain commodities (like rice and beef), and providing welfare services.[53] Public enterprises thus do much of the nuts and bolts work connected with the implementation of Japanese industrial policy. It is hard to imagine how Japanese industrial policy would be carried out if MITI did not have the authority to organize such things as cooperative research associations for its national research projects; one example is the Information Technology Promotion Agency (IPA), created for the coordination of software development programs.

The network of formal intermediate organizations can be conceived of as the arms and legs of the state. In some senses—without carrying the comparison too far—the government–public enterprise relationship is not unlike the relationship between large parent corporations and their cluster of affiliated companies, subsidiaries, and subcontractors. Just as small, new subsidiaries are created as alternatives to internal diversification and expansion, public enterprises and special legal entities can be organized to meet the growing and ever-changing functional needs of public administration. The intermediate zone—the realm of public corporations and mixed public-private enterprise—is expandable, like the second tier of Japan's dual industrial structure.

Perhaps it would not be altogether inaccurate to characterize the relationship between state bureaucracies and intermediate organizations as Japan's dual administrative structure, in which the latter—the second tier—serve as a bridge between the public and private sectors. MITI has 27 public enterprises under its jurisdiction, the largest number of any ministry, and this does not include the scores of special legal entities also attached to it. Some of the same considerations are true here as in the case of Japan's structure of corporations and subsidiaries. For example, there is a strong proclivity for government ministries to protect their lifetime labor force from accordion-like expansions and contractions and to keep the scope of their work clearly focused. By establishing special legal entities, the government ministries can manage the growing load of administration, create new positions for retiring officials, and extend their influence over new functional domains without incurring the higher fixed costs of internal expansion.

Recognizing that the government consists not simply of central ministries but also of the much bigger but less visible stratum of intermediate organizations again helps to cast clearer light on the seeming paradox of Japan as both a minimalist and an interventionist state.

Looking only at central institutions, it would be easy to conclude that the minimalist image is the most accurate, especially if one overlooks the state's latent capacity to intervene selectively through multiple points of access. Certain issues can be handled at the central level without utilizing institutions located in the intermediate zone.

If, on the other hand, the focus is broadened to encompass the maze of intermediate organizations, the interventionist image might seem closer to reality. Most issues, which cannot be handled by MITI alone, require the involvement of intermediate organizations below the level of the central government. The maze of quasi-governmental organizations and informal policy networks (discussed below) give the state enormous extension into the hinterland of Japan's political economy. Which image best fits reality—the minimalist or maximalist state—therefore depends on which administrative structure is activated: the central bureaucracies alone or the combination of ministries and intermediate organizations. The perspective provided by the notion of a dual administrative structure—including central and intermediate organizations, formal and informal channels, purely public and mixed public-private groups—thus sheds light on the complex but comparatively flexible machinery of the Japanese state.

In addition to quasi-governmental and mixed public-private organizations, policy networks also include informal relationships between government officials and industrial leaders. From the standpoint of the state's capacity to influence Japan's industrial economy, informal policy networks may be at least as important as public corporations. Informal networks give ministries like MITI the latitude to discuss problems, work out differences, and build consensus with the private sector.

If its officials did not devote so much time and energy to cultivating personal relationships with key leaders in the private sector, MITI would not be nearly as effective as it is. Indeed, early consultation, on-going negotiation, conflict resolution, and consensus formation— the sine qua non of effective industrial policy and close government-business relations—hinge, to a large extent, on the existence and utilization of informal networks. Similarly, MITI's capacity to handle the enormous demands of industrial policy-making with the limited resources at its disposal hinges on its ability to transfer some of the burdens of policy-making from its own corridors to the maze of intermediate organizations and informal policy networks, where much of the time-consuming work of consensus formation takes place.

Informal policy networks, as one might expect, grow mainly out of work-related contacts between government officials and corporate leaders. But they are a product of both functional roles and ascriptive affiliations. Included under ascriptive networks are marital and kinship relations (*keibatsu*), common place of origin, contact through mutual friends, school ties (*gakubatsu*), especially graduation from the University of Tokyo, and school club ties. Functional ties emerge from friendships developed in the course of government-industry contact, participation in informal study groups, and amakudari networks. Several factors have conspired to consolidate the importance of informal policy networks in Japan: a "frame" society based on consensus rather than legal codes; the Confucian stress placed on human relations, particularly the emphasis on loyalty and trust; social homogeneity; the logistical convenience of Tokyo as the hub of both political and economic activity; and the role of the educational system as the central mechanism for elite selection and social mobility. This combination of factors appears to make Japan perfectly suited to the informal, non-legalistic policy-making that is a hallmark of its administrative apparatus.

Japanese policy networks are extensive and serve as channels for the transmission of valuable information. Business executives are often willing to take MITI officials into their confidence, sharing sensitive information, even to the extent of divulging what may be proprietary in nature. This is done, as pointed out in Chapter 2, on the basis of reciprocal, confidential disclosure. Each side is expected to give information of roughly equivalent value to the information that is received and to keep what is received strictly confidential.

This intimacy of interaction is possible in Japan for several reasons: (1) MITI officials are viewed as impartial and supportive; (2) there is no tradition of ideological bias against public-private interpenetration; (3) government and industry share overriding interests; (4) most industries look to MITI for some kind of support; (5) the private sector, on balance, has been satisfied with MITI's industrial policy; and (6) the advantages of long-term cooperation are perceived to outweigh any incentives for industry to engage in behavior that Oliver Williamson calls "opportunistic" or "self-seeking with guile."[54] Such openness is hard to envision in the United States; although one or more of the necessary conditions may exist there, the combination of factors does not.

There are, of course, certain pitfalls in relying on informal policy networks. Falling captive to specific industrial interests is one. Owing to the nexus of informal ties with industrial leaders, MITI officials in the vertical bureaus (genkyoku) and divisions (*genka*) cannot help developing a strong sympathy for the interests and needs of industries under their jurisdiction. There is a built-in tendency for them to advocate policy measures that advance those interests and to see them as contributing to the collective good. So long as industry-specific interests are relatively self-contained, do not impinge on those of higher-priority sectors, and do not otherwise harm the economy, they can pass through MITI's policy-making channels without opposition or fundamental revision. However, because MITI's horizontal (that is, functional) divisions must find ways of integrating policies advocated by all genkyoku units, the needs and interests of one industry have to be balanced against those of others. As discussed earlier, this provides built-in safeguards against responding politically to industrial demands as if they are short-run, variable-sum situations when they are often, in reality, long-run, zero-sum propositions.

For deputy directors of genkyoku, or vertical divisions, the most important duty is to stay in close touch with business leaders from industries under their supervision. It is not at all unusual for MITI deputy directors to spend more than half their time with key personnel from large corporations (especially directors of research and development), industrial association (*gyōkai*) officials, and members of other private-sector groups.[55] Regular interaction is indispensable for developing the kind of binding trust on which MITI-industry relations rest.

Over the course of repeated contact, during which the boundaries between work-related interaction and personal socializing become blurred, the relationship between MITI bureaucrats and industrialists often moves beyond what is purely business into the realm of personal friendship. Just as the peculiar blend of market and organization sets Japanese industrial organization apart from that of other advanced countries and explains the distinctive behavior of Japanese corporations, so too the blurring together of professional and personal ties gives government-business relations their distinctively Japanese flavor. This phenomenon might be described as the "personalization" of professional interactions, or the intrusion of affective bonds into the domain of public policy-making. As a result, the lines of demarcation separating government from industry lose the sharpness retained in the United States.

The fusion of what tend to be regarded as mutually exclusive opposites—market and organization, public and private, and formal and informal—stands out as a notable characteristic of government-business relations in Japan. Of course, some degree of commingling is observable wherever market conditions exist. What gives Japan's political economy its distinctive cast is the peculiar composition and extensiveness of the interpenetration. Although Japanese industrial policy is basically market-conforming, the private sector operates within the framework of institutions that shape market forces and permit MITI to fine-tune the functioning of the industrial economy without having to wield too heavy a hand.

With the exception of certain sectors like agriculture and defense, government-business relations in the United States, by contrast, can be characterized as formal, distant, rigid, suspicious, legalistic, narrow, and short-term-oriented. Underlying this strained relationship is the divergence—often a frontal clash—between private interests and the public good.[56] Thanks in part to the interpenetration of public and private domains and to MITI's ability to reconcile private interests with the public good, government-business relations in Japan are informal, close, cooperative, flexible, reciprocal, nonlitigious, and long-term in orientation. Their nonantagonistic nature is a major asset and helps to explain why the administration of Japanese industrial policy has been relatively smooth.

Compared to other bureaucracies, MITI is unusual in that it gives younger officials the opportunity to demonstrate initiative in defining and carrying out their work responsibilities. Unlike their counterparts in, say, the Ministry of Foreign Affairs, deputy division directors in MITI's vertical divisions have significant leeway to structure their relationships with corporate executives in ways that fulfill their notions of what their jobs should entail. Usually, there is a fairly well defined hierarchy of corporations, based on size, market share, and prestige, that defines the universe of companies with which deputy directors must deal. The three deputy directors of the information industry divisions—Electronic Policy, Data Processing Promotion, and Industrial Electronics—are expected to establish close working relations with representatives from the five big, blue-chip electronics corporations that dominate their fields—Hitachi, NEC, Toshiba, Mitsubishi Electric, and Fujitsu (see Fig. 3.5).

Working with industrial associations like the Electronics Industry Association of Japan (EIAJ) is also essential, especially when the association is strong or when there are too many large corporations for one or two MITI officials to handle. From the deputy division director's standpoint, staying in close touch with industrial associations is administratively convenient because of the association's role as a central communication link to nearly all private corporations in its in-

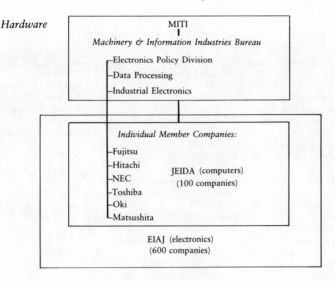

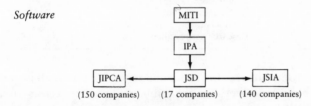

Fig. 3.5. Informal policy networks

dustry. The involvement of leading corporations and industrial associations and their collective capacity to work out intramural differences alleviates the burden of aggregating the diverse, sometimes conflicting, array of private-sector demands.

For political-economic systems like Japan's that function on the
basis of consensus and not in accordance with binding rules and laws,
the role of informal policy networks for industrial policy-making can
scarcely be overstated. Such networks serve as the main vehicle for the
painstaking rounds of informal consultations (*nemawashi*) that are the
sine qua non of consensus building. To function effectively, therefore,
Japan's consensus-oriented system requires the full utilization of the
informal policy networks that occupy the intermediate zone between
MITI and the marketplace (see Table 3.6).

Of course, Japan is not the only place where the lines of demarca-

TABLE 3.6

Intermediate Zone Networks

Government	Intermediate zone	Private industry
MITI	Industrial associations	Hitachi
Machinery and	EIAJ	Nippon Electric (NEC)
Information	JEIDA	Toshiba
Industries Bureau	Others	Mitsubishi Electric
Electronics Policy	Informal study groups	Fujitsu
Division	School and family ties	Other companies
Data Processing	Amakudari connections	
Promotion	LDP Diet-member caucuses	
Division	Advisory councils	
Electronics and	Public corporations (*kōsha*)	
Electrical	Public enterprise corporations	
Machinery	(*kōdan*)	
Division	Enterprise corporations	
Informal policy	(*jigyōdan*)	
networks	Public finance corporations	
Formal public	(*kōko*)	
policy	Mixed public-private	
companies	companies (*tokushu kaisha*)	
	Foundations (*kikin*)	
	Industrial promotion	
	associations	
	Research associations	

tion have become blurred. In France and several of the small European states, the boundaries between the public and private sectors are also hazy. Furthermore, the structure of neocorporatism, so pervasive in Western Europe, also appears to have broken down traditional lines of demarcation between the state and private enterprise.[57]

This pattern of interpenetration even exists in select pockets of the U.S. political economy: specifically, between DOD and the defense industry (the so-called military-industrial complex), between NASA and the aerospace industry,[58] and between the Department of Agriculture and the farming community. In all three cases, the relationship is held together by compelling national security concerns, a close identity of interests, mutual dependence, lateral mobility of personnel, informal policy networks, and huge procurement budgets. Not accidentally, the three happen to be the most striking examples of government-business cooperation in the United States. The fact that they are the exceptions rather than the rule, and that cooperation is conspicuously missing elsewhere in the U.S. economy, implies that public-private interpenetration and dense informal networks may be a necessary but not sufficient condition for close government-business relations.

Of course, the mere existence of informal networks is no guarantee that there will be cooperation or effective industrial policy output; examples of government-industry friction, ineffectual industrial policy, and economic inefficiency abound in countries where there is considerable public sector–private sector overlap.[59] Moreover, such intervening variables as the relative strength of organized labor can also make causal inferences questionable. Nevertheless, because so few examples of cooperation *sans* policy networks can be cited, there seems to be at least a crude correlation between the structure of public-private interpenetration and the potential for government-business cooperation.

In Japan, the composition of MITI-industry networks continually changes. Newly appointed deputy directors are forced periodically to expand or reconstitute their networks because of regular turnover in personnel. A skillful deputy director might succeed in establishing relations with key business leaders who, by virtue of their stature, force of personality, or company power, are able to galvanize intra-industry consensus. Needless to say, having one or two such allies in the private sector can be of immense value in working out complicated policy issues. Indeed, the deputy division director's job performance hinges to a large extent on the effectiveness of the human networks (*jinmyaku*) he has been able to develop.

Amakudari: Interpenetration of Personnel

The interpenetration of public and private is nowhere more graphically illustrated than in the well-known practice called amakudari, whereby officials leaving the bureaucracy "descend from heaven" into high-level posts in public corporations, industrial associations, and private industry. According to the annual report published by the National Personnel Authority, 159 bureaucrats accepted positions in the private sector during 1976.[60] The largest group, 44 strong, "descended" from the Ministry of Finance; MITI, along with the Ministry of Transportation, ranked second, with 15, or only one-third the number sent by MOF. At the bottom of the list, with only one placement, was the Ministry of Foreign Affairs, a prestigious bureaucracy but one without a clearly defined domestic interest group under its jurisdiction; alone among Japan's central bureaucracies, it can guarantee regular promotion for all its higher civil servants into ambassadorial posts around the world.

On the demand side, the industries that hired the largest number of bureaucrats were construction (42), finance (34), and transportation (28). In the high-technology sector, electronics and electrical machinery—a broad category that includes manufacturers of consumer and industrial electronics—employed only 8 of the 159 amakudari bureaucrats, or about 5 percent of the total. Of the eight, only three—two administrative officials (*jimukan*) and one technical specialist (*gikan*), came from MITI. Only one landed a job in a major blue-chip corporation (NEC).

The number of ex-MITI bureaucrats entering the electronics industry is growing, however, as this sector matures and becomes the driving force behind the development of Japan's postindustrial economy. Such is the pattern of amakudari placement: it tends to reflect trends in the leading growth sectors of the economy. During the 1960's, the highest-ranking MITI officials tended to "parachute" into what was then Japan's most important sector, steel. Since the beginning of the 1980's, however, MITI's most outstanding bureaucrats have moved in growing numbers into the information industry because it is the leading strategic sector for Japan's transition from a heavy manufacturing–based to a high technology–based economy. MITI officials tend to want to go to industries where the growth prospects appear brightest. Thus, the amakudari system winds up sending retired MITI officials to the very sectors identified as most central to the development of Japan's economy.

Lateral mobility on a sizable scale is also observable in other countries, like France, and in certain sectors of the United States. The traffic of personnel is heavy among positions in DOD, NASA, and defense-aerospace companies. From 1974 to 1979, more than 75 employees moved between NASA and eight large aerospace companies; from 1970 to 1979, more than 1,675 moved between DOD and the eight defense contractors. The importance of R&D contracts and procurements is reflected in the large number—over one-third of those transferred—assuming positions directly related to R&D activities.[61] Here again, the fact that a revolving door operates in one of the few sectors in the United States characterized by close government-business cooperation is probably not coincidental. It suggests that the interchangeability of key personnel lubricates interactions between the public and private sectors and consolidates cooperative relations (albeit with discernible costs and dangers).

The system of amakudari serves a variety of functions.[62] For government ministries, it provides attractive landing spots for high-level officials who fail to rise up the narrowing pyramid of the bureaucratic hierarchy. By placing retiring bureaucrats in lucrative posts in the private sector, MITI is able to reward officials for years of hard work and loyal service. To some extent, amakudari provides compensation for years of wage disparity between the public and private sectors while at the same time ensuring that MITI careers continue to be attractive to the best and brightest college graduates. The system of amakudari also enhances MITI's influence by placing its alumni in key positions within many of the country's strategic industries and public corporations. This network of strategically placed alumni can be used to build consensus on issues of industrial policy.

The benefits for private corporations are similarly significant. The system of amakudari gives private firms access to men of exceptional talent and experience, former bureaucrats who understand firsthand how the policy-making processes work and who maintain close, personal ties with former MITI subordinates and colleagues. Such contacts can be of substantial value in terms of (1) finding out where MITI stands with respect to public policy questions; (2) having an input in industrial policy-making; (3) obtaining approval for a variety of activities requiring governmental authorization (such as sites for new plant facilities); (4) soliciting MITI support for R&D subsidies, national research projects, tax policies, administrative guidance, and various pieces of legislation; and (5) gaining MITI's assistance on matters related to international trade and investment. Private companies in a number of industries still look to MITI for support across this spectrum of activities. The system of amakudari thus provides powerful reinforcement for the structure of ties linking government and business.

Less widely publicized but just as important is the movement of ex-bureaucrats into a variety of quasi-governmental organizations, including public corporations, industrial associations, special legal entities, and nonprofit foundations—all located somewhere at the mezzanine level between the higher civil service and the private sector. In the industrial associations representing Japan's information industries, for example, former MITI bureaucrats occupy key administrative posts; they work in the Electronics Industry Association of Japan (EIAJ), Japan Electronics Industry Development Association (JEIDA), Japan Information Processing Development Center (JIPDEC), Japan Software Industry Association (JSIA), and Japan Information Processing Center Association (JIPCA). Former MITI officials can also be found in the Information Technology Promotion Agency, a semigovernmental organization that plays a central role in the development of Japan's software and data processing industries. Although the posts are not particularly powerful or prestigious, they provide MITI with useful communication links to key sectors. MITI sometimes pays the salaries of former officials in semigovernmental agencies by drawing on revenues from taxes on legalized gambling in bicycle racing (*keirin shikin*). Here again, the transplantation of personnel strengthens the structure of intermediate organizations while at the same time extending the radius of MITI's influence and consolidating its base for industrial policy-making.

The system of amakudari has been roundly criticized by opposition parties and progressive intellectuals for promoting incestuous inbreeding and structural corruption.[63] Quite apart from the dangers of overt malfeasance, the practice of amakudari poses disquieting opportunities for potential conflicts of interest and abuses of power.[64] The concern is that MITI officials will be tempted to curry favor with specific companies in order to line up post-retirement jobs. Even if the temptation is resisted, the structure of the situation is such that MITI officials may find themselves under subtle but persistent pressure to devise policies highly favorable to the interests of particular companies or industries.

On the basis of these fears, certain laws have been passed to counteract the potential conflicts of interest. Article 103 of the National Public Service Law prohibits civil servants from moving immediately into companies with which they have worked prior to their retirement from government service. Amakudari bureaucrats must wait two years in "purgatory"—temporary positions in unrelated businesses or in nonprofit organizations—before entering private companies to begin their post-MITI careers. Moreover, retiring bureaucrats are not given the freedom to choose which company or nonprofit organization to enter. Companies approach MITI about job openings, asking perhaps that a particular individual whom they trust be assigned to them. The Minister's Secretariat plays the role of intermediary, matching the incoming requests from companies with the credentials and expressed desires of the retiring bureaucrats. Under this system, the amakudari bureaucrat might not be placed with the company, or indeed the industry, of his choice.

This suggests that the practice of amakudari does not automatically strengthen MITI's power vis-à-vis the private sector. The number of ex-MITI officials in the electronics industry, as pointed out earlier, is still relatively small. Although the number is increasing rapidly as the industry itself grows and becomes increasingly central to Japan's postindustrial economy, the total number of former higher civil servants from MITI (*jimukan*) among the eight leading corporations has been estimated at less than twenty (in 1988).[65] Moreover, amakudari bureaucrats are bound to encounter resentment when they enter corporations as "outsiders," who have not paid their dues like everyone else. To overcome the stigma of being a privileged outsider they sometimes seek to gain credibility with their company colleagues by avoiding activities that involve direct liaison with MITI. They may also try to assimilate company values quickly.

From its own perspective, too, MITI is careful not to overuse its connections with amakudari bureaucrats. MITI realizes that the second careers of its ex-officials can be compromised if they come to be regarded simply as in-house lobbyists, communication couriers, or, worst of all, planted spies. This means that elitist or conspiratorial interpretations of amakudari—as MITI's Trojan horse—should be taken with more than a grain of salt. One should not exaggerate the scope of the influence gained by MITI through the system of amakudari. More than expanding MITI's own power, the system works to consolidate the structure of government-business ties and the interpenetration of the public and private sectors through the network of intermediate organizations.

Industrial Associations

In Japan, where the challenge of organizing for collective industrial or sector-specific interests has been achieved with greater frequency and success than in the United States or most European states, industrial associations (IAs) have played a big role in aggregating individual company interests, building intra-industry consensus, and serving as a vehicle of communication between industry and government. Of course, the strength of IAs varies across sectors according to such factors as the number of member companies and the degree of market concentration. Mature, basic industries with a small handful of producers, heavy capital investments, limited leeway for product diversification and technological expansion, and a high degree of dependence on MITI tend to have strong IAs. Rapidly growing industries, on the other hand, with a large number of companies, greater

potential for technological change, wider leeway for product diversification, and less reliance on MITI tend to have less cohesive IAs. The
spectrum in Japan is broad, ranging from strong IAs in basic sectors
like steel and electrical power utilities to weaker and more loosely
organized ones in high-tech sectors like electronics and the information industries.

From the standpoint of the policy-making processes, the stronger
the IA, the lighter the burden that falls on MITI, assuming the industrial association and MITI agree on basic policy issues. An active and
strong IA can greatly facilitate the process of consensus formation.
From MITI's perspective, the advantages of working with and through
strong IAs can be substantial, though there are potential costs, too,
particularly if the industrial associations and MITI cannot reach
agreement. Even though IAs themselves may not be independently
powerful, they can still serve a useful function as mechanisms of centralized communication, linking MITI to specific sectors of the
economy.

The role of IAs draws attention to an extraordinary feature of Japanese industry: namely, the balance it somehow manages to strike
between fierce interfirm competition and a willingness to cooperate on
basic matters perceived to lie in the common interest. In few other
countries is this seen to the same degree. When private corporations
can overcome the classic "prisoners' dilemma" and work together to
realize collective interests—in cooperation with MITI—the net effect
is to insulate industrial policy from the kind of parochialism and
politicization that take place in the United States when interest groups
attempt to manipulate the Congress in ways that maximize their own
narrow, short-term interests. Japan's capacity to forge consensus,
based on extensive communications within industry and between government and industry, is one of the main reasons that it has succeeded
where other industrial states have failed in protecting industrial policy
from excessive politicization.

The Electronics Industry Association of Japan

Having pointed out the advantages of MITI-IA cooperation, I now
examine how MITI has interacted with Japan's biggest IA in the field
of electronics, the Electronics Industry Association of Japan (EIAJ). In
spite of the fact that an ex-MITI man currently serves as executive
director of EIAJ, MITI's relationship with EIAJ has been uneasy, at
times strained, and not nearly as cooperative as its relationship with
other IAs. The causes of MITI-EIAJ tensions are many. They can be

traced to the 1970's, when, in response to strong U.S. pressures, MITI used its powers of persuasion to induce the electronics industry to accept voluntary export restraints (VERs) on color televisions and video tape recorders (VTRs). Member companies, and the EIAJ itself, were unhappy with the way MITI caved in to U.S. demands, forcing Japanese producers to swallow quantitative limits on exports. They felt as if the electronics industry had been made the sacrificial lamb to appease the anger of U.S. industry and maintain harmony in the United States–Japan alliance.

Personality conflicts involving EIAJ's executive director and certain MITI officials no doubt have ruffled relations, too. Though he is a MITI man, the executive director took industry's point of view and openly criticized MITI for its handling of the trade negotiations. As a result, the mutual trust and open lines of communication associated with MITI's relations with the Japan Iron and Steel Federation, for example, failed to develop. There are times, in fact, when EIAJ not only neglects to consult with MITI but actually tries to bypass MITI and reach decisions on its own, without MITI's prior approval or involvement. A change in the personal chemistry of policymakers on both sides might soothe ruffled feelings, but the sources of strain run deep.

Other problems involve structural factors. Over 600 companies of various sizes and types, in vigorous competition with each other, belong to EIAJ; this makes genuine consensus on policy issues difficult. Moreover, member companies produce everything from semiconductors to computers and consumer electronics, creating conditions that can easily lead to internal rifts and divergent perceptions of what lies in the collective interest. Take, for example, two products, semiconductors and consumer electronics, that represent two different divisions within the same corporation. Dependent on MITI for R&D subsidies, the semiconductor divisions of large Japanese corporations tend to take a cooperative attitude, keeping in close touch with MITI through formal channels of communication and informal policy networks. The consumer electronics divisions of the same companies, relying relatively little on MITI, have a more distant and independent relationship (against the background of conflict caused by the issue of voluntary restraints on exports to the United States). With so many member companies and diverse interests to deal with, EIAJ has not been able to develop the cohesion or unity of purpose necessary to work in close concert with MITI.

The EIAJ-MITI relationship brings to light something that is often taken for granted: namely, that Japanese government-industry cooperation rests on a harmonization of public and private interests, or what Charles L. Schultze calls the government's capacity to harness the tremendous energy of private enterprise for the achievement of public goals, utilizing (instead of neutralizing) the strength of private incentives.[66] Given MITI's promotional approach to industrial policy, the interests and objectives of the two sides usually dovetail. In cases where they do not, a concerted effort is made by both sides to negotiate a compromise. As part of the implicit rules and norms governing the system of consensus, private corporations are usually willing to hold up their end of the relationship by meeting the government halfway; the same holds true for the government. In the unusual event that a satisfactory compromise cannot be worked out, the result can be general paralysis or an attempt by MITI to impose a policy solution that the private sector is apt to resist or hold against MITI in subsequent negotiations.[67]

Tensions between EIAJ and MITI, coupled with the problems of aggregating member company interests, reveal the basic fallacy of the "Japan, Inc." stereotype, especially when applied to government-business relations in high technology. Japanese electronics companies are too large, diversified, and autonomous for MITI to control. Instead of forcefully imposing its will on this rapidly growing industry, MITI has wisely sought to convert the industry's dynamism into progress toward public goals by designing policies that cater to private incentives for growth. The underlying assumption is that the growth of the information industries advances the public good and serves the national interest.

This is not to say that MITI's relationship with the electronics industry is so difficult as to be unmanageable. It is simply not as close as the relationship with steel, petroleum refining, or the electrical power utilities—the basic, raw-materials-converting industries whose particular characteristics, as pointed out earlier, make them more dependent on MITI. But neither is it as distant as the relationship with producers of precision machinery, the fast-foods industry, or the general trading companies. It is somewhere between adhesion (*yuchaku*) and arm's-length autonomy.

On broad, long-term objectives, such as desirable future technological directions, MITI and the information industries have always managed to reach consensus. Agreement is much harder to reach when it comes to specific, short-term goals, such as production levels or product development—matters best left to the judgment of private corporations. This not-too-close yet not-too-distant relationship, based on MITI's willingness to bow to the judgment of private enterprise, is entirely in keeping with the ever-changing requisites of industrial life cycles. For industries at an adolescent stage, with developmental potential, the in-between relationship appears to be well suited to promoting efficiency and growth.

Leaving the processes of industrial maturation to the marketplace means that EIAJ-MITI frictions pose less of a problem than would be the case if conflicts of similar severity arose between MITI and, say, the basic raw-materials-converting sectors. Where high technology is concerned, MITI does not have to rely as heavily on IAs to aggregate interests or to serve as centralized switchboards of communication. This is the beauty of Japan's spacious intermediate zone. There is plenty of slack for MITI deputy division directors to set up their own informal networks with key leaders from major corporations. It may take time to create and maintain these unofficial networks, but the job can be done at marginal incremental costs. Such informal networks permit MITI the flexibility to find ways of working with private industry when it cannot rely on IAs.

EIAJ is also not the only IA in the information industries sector. Because the electronics industry is so diversified, a large number of business associations have had to be organized across a broad spectrum of products. These include JEIDA, with 100 member companies, for computers and JSIA, with 140 member companies, and JIPCA, with 150 member companies, for software. As in the case of EIAJ, ex-MITI officials hold key posts within each of these associations; however, MITI's relationship with each is cordial, a fact that has helped to offset the opportunity costs incurred as a result of its strained relationship with EIAJ.

In what Chie Nakane has labeled Japan's "frame society," the strength of particularistic, vertical organizations and the relative weakness of horizontal associations have created both the need and the opportunity for MITI officials to develop extended policy networks with key actors in private industry, transcending membership in individual organizations, and linking the public and private sectors together in an intermediate zone of contact. MITI officials can forge consensus with industry and implement industrial policy even when its relationship with one of the key IAs is less than cordial. As guardian of the national interest, MITI is poised to play a pivotal role in consensus formation and resource mobilization.

The coordinating role is especially important in the Japanese political economy because of the binding sense of particularistic loyalty that comes with career-long employment and the fierce competition that pits one corporation against another. In a system organized around the dynamism of private corporations, the forces of market competition work against efforts at horizontal cooperation.[68] As Nakane observes, "The entire society is a sort of aggregation of numerous independent competing groups which of themselves can make no links with each other."[69] Although Nakane overstates the weakness of horizontal organizations while underestimating the potential for building cross-cutting linkages, she accurately diagnoses Japan's central tendencies and correctly calls attention to the need for mechanisms of horizontal communication and consensus formation.

Nakane believes that Japanese society needs centralized coordination from above in the form of political authority to hold centrifugal forces in check and keep society from pulling apart. The bureaucratic state is able to fill this need precisely because the vertical structure of Japanese society weakens the private sector's capacity to challenge government's authority:

These characteristics of Japanese society assist the development of the state political organization. Competing clusters, in view of the difficulty of reaching agreement or consensus between clusters, have a diminished authority in dealings with the state administration. Competition and hostile relations between the civil powers facilitate the acceptance of state power and, in that the group is organized vertically, once the state's administrative authority is accepted, it can be transmitted without obstruction down the vertical line of a group's internal organization. In this way, the administrative web is woven more thoroughly into Japanese society than perhaps any other in the world.[70]

Such Japanese values are oriented toward the goal of achieving the good of the group or collectivity, and since the nation is the ultimate collectivity, the authority vested in state carries far more weight than that merely stipulated in the provisions of the constitution. To the state falls the task of harmonizing the collage of interests articulated by corporate and organized lobbyist groups and blending them in the pursuit of broad national interests. Here again, though Nakane fails to go into it, Japan's extensive reliance on informal networks can be regarded as the sine qua non not only of effective state power but also of the capacity of both government and industry to fine-tune industrial policies to meet the ever-changing needs of the Japanese political economy.

Business Federations: Keidanren

The thoroughness with which Japanese business is organized is reflected not only in sector-specific industrial associations like the EIAJ but also in comprehensive business federations like Keidanren (the Federation of Economic Organizations), Keizai Dōyūkai, and Nisshō (Japan Chamber of Commerce). These federations bring together business leaders from nearly all segments of the manufacturing, financial, and service sectors. Such broad-based associations offer central forums for the exchange of information and regular policy deliberation. By examining business and economic issues from a broad, cross-sectoral point of view, executive leaders are able to influence one another, work out differences of opinion, and advocate positions that transcend narrow company- or industry-specific interests. Business federations stand at the apex of the structure and processes of private-sector interest aggregation—from the individual firm to the banking-business nexus, industrial associations, keiretsu groupings, and inclusionary business federations. At each level of aggregation, the business perspective tends to broaden.

Although the inclusionary business federations have prestige, the power they wield is often blown out of proportion.[71] Keidanren's relationship with MITI is not nearly as close or collaborative as some writers intimate. As a rule, Keidanren refrains from taking stands on industry-specific issues, which it leaves to MITI and specific industries to sort out. It confines itself instead to macroeconomic and societal issues, such as money supply, interest rates, exchange rates, taxes, and so forth. This means that it interfaces less with MITI than with MOF and the Bank of Japan. Although big business is a key part of the ruling coalition, inclusionary associations like Keidanren and Keizai Dōyūkai play a peripheral role in the formulation of industrial policy. In spite of its international notoriety, Keidanren takes a back seat to industry-specific actors—leading companies, industrial associations, and informal policy networks.

Even with respect to macroeconomic issues on which it is not hesitant to speak out, such as monetary policy, Keidanren's voice is only one of many. When the monetary authorities restrict money supply for sustained periods in order to curb inflation, Keidanren can be counted on to speak out in favor of loosening the reins in order to stimulate business demand; but seldom is its weight sufficient by itself to sway

the Bank of Japan or the Ministry of Finance. Nor, despite its overwhelming stake in free trade, has Keidanren been able to force farmers
to lift import barriers. Indeed, when Keidanren's subcommittee on
agriculture criticized farmers for resisting liberalization, its report
prompted the 78,000-member Hokkaido Agricultural Cooperative to
boycott companies whose chief executive officers had criticized agricultural protectionism. This forced the head of the Keidanren committee to apologize and resign from the chairmanship of the committee. The episode indicates that Keidanren and other inclusionary
associations are not as influential as the impressive roster of member
companies might lead one to expect.

Japanese Industrial Organization and Industrial Policy

This chapter has argued that one of the "secrets" to the comparative
effectiveness of Japanese industrial policy can be found in the characteristics of Japanese industrial organization and government-business
cooperation that emerges from it. Industrial policy cannot be understood as merely an abstract set of policy measures, detached from the
political-economical context within which it is administered. The institutional context not only shapes the substance of industrial policy
but also determines how it is implemented. The interface between the
public and private sectors has given MITI the opportunity to devise a
very broad range of policy measures, fine-tuned to meet the needs of
specific industries, without having to resort to costly and inefficient
measures that ultimately undermine industrial competitiveness. What
Japanese industrial organization and the structure of government-
business relations provide is the leeway for MITI to intervene only
when necessary, thanks to the availability of structural "handles" for
government coordination.

In this chapter, three spheres of industrial organization and government-business ties have been analyzed: (1) bureaucratic institutions, especially the scope and organization of MITI, and the regulated
financial system; (2) Japan's distinctive set of organizational characteristics, including the relative weakness of organized labor, the extensive system of subcontracting, and keiretsu groupings; and (3) the
role of intermediate organizations such as public corporations, industrial associations, business federations, and informal policy networks.
The three spheres serve as pillars for what might be called a dual
administrative structure that supports a lean government at the top

while extending its reach deep into the core of the expansive private sector when intervention is necessary.

Of the various elements that constitute Japan's institutional structure, which are commonly found in other advanced industrial states and which are distinctive to Japan? Sorting out the common from the distinctive should help us clarify exactly what makes Japan different. Let us identify, first, the common features.

Characteristics Shared with Other Countries

A strong central bureaucracy (France, England)
  Superior quality, elite officials
  Established routes of recruitment through an educational hierarchy
  Retirement into top-level positions in public/private sectors (amakudari)
Extensive financial regulation (France)
  High debt-to-equity ratios in corporate financing
  Some government influence over credit allocation (rapidly waning in Japan)
  Close banking-business relations (West Germany)
Blurred boundaries between public and private sectors (France, Italy)
  Quasi-governmental organizations (U.S., Western Europe)
  Informal policy networks (France, some segments of U.S.)
  Intersectoral mobility (U.S., France)
  Industrial associations (France, U.S.)
  Advisory councils
High market concentrations (France, Italy)
*Subcontracting (U.S., several Western European states)
*Company specialization
*Keiretsu groupings (South Korea)

*Denotes areas where there are significant differences in degree in spite of surface similarities.

The list suggests that certain institutions considered unique to Japan, such as amakudari, can be found in other industrial countries. Some institutions, like subcontracting, take different forms; in Japan the relationship is tighter and based on longer-term commitments and reciprocal obligation. Others, like company specialization, differ only in degree, not in kind. But the underlying similarities are sufficiently strong to remove them from any list of distinctively Japanese attributes. Characteristics that fall into the latter category are listed below:

Distinctively Japanese Characteristics

MITI and its internal organization
  Organization into vertical (industry-specific) and horizontal (functional or cross-industrial) axes
  A bottom-up mode of policy-making, with power concentrated at the level of deputy division director
  Extraordinarily broad jurisdiction encompassing everything from small and medium-sized enterprises to petroleum refining
Comparative insulation from dictates of capital markets
  Disavowal of short-term profit maximization strategy for corporations
  Low return (comparatively) on assets and return-on-investment pressures
Japanese labor
  Career-long employment at large corporations
  Company-centered, enterprise labor unions
  Flexibility through part-time and female labor force
  Comparative weakness of organized labor in national politics
Relational state: mutual dependence of state and society
  General convergence of private interests and public goals
  Government's sustained probusiness orientation
  Consensus system
  Government-business relations based on long-term trust and mutual dependence
  Importance of informal policy networks and intermediate organizations
Far-reaching interpenetration of market and organization
  Long-term business relationships based on market signals, intercorporate stockholding, and reciprocal obligation
  Reduced opportunities for exit; more stress on voice and loyalty
  Company-centered, group-oriented competition (less emphasis on individual incentives and interests)
  Blending of meritocratic and ascriptive criteria within Japanese organizations
Society and values
  Non-legalistic, relational society
  Voluntaristic behavior; limited reliance on coercive power
  Strength of vertical, "frame" organizations

Work ethic, achievement orientation, and Confucian value system
Consciousness of national identity
Interaction and combination of all the above

The last point deserves underscoring. Japan's distinctiveness does not lie in the uniqueness of its institutions, be they keiretsu groupings or intermediate organizations; counterparts or functional equivalents can be found in other countries. What sets Japan apart from other political economies is the blend of all the characteristics listed above. It is the dynamic, interactive chemistry of the whole that gives the Japanese political economy its distinctive cast.

Looking at Japan's system from the standpoint of policy outputs—specifically industrial policy—perhaps the seven factors in the second list can be distilled into two overriding features: (1) the intrusiveness of organizational factors in Japan's market economy, especially its capital and labor markets; and (2) the compatibility of Japan's particular blend of market and organization with the practices of government-business consensus, government-led coordination, and selective intervention. It would be hard to find an institutional structure better suited for the implementation of industrial policy.

Because private enterprise is internationally competitive in most manufacturing sectors, the government is spared the burden of costly protectionism and subsidization in declining sectors. The private sector is also capable of overcoming the prisoners' dilemma as it pursues corporate and collective goals. The capacity of the public and private sectors to reach consensus and work together toward the achievement of common objectives is one of the keys to Japan's ability to protect industrial policy from political ambush. What Japan's institutional structure has offered the state is a network of access points through which to shape market forces, intervene selectively when necessary, and pull back afterward. This averts the "ratchet effect" of an ever-expanding administrative structure and government role.

Of course, the explosive growth of Japan's economy has tilted the balance of power between the public and private sectors in the private sector's direction, complicating the processes of industrial policymaking. With Japan's emergence as the world's second-largest economy, international pressures have also been stepped up, forcing Japan to make changes in its domestically oriented system of industrial policy. These developments have altered the relationship between bureaucrats and businessmen—not to mention that between bureaucrats and politicians. Having analyzed Japanese industrial organization and the linkages between the public and private sectors, let us turn to an examination of the political system and the interactions between bureaucracies, the ruling conservative party, and key interest groups.

FOUR

The Politics of Japanese Industrial Policy

In this chapter, the relationship between political regime characteristics and Japanese industrial policy will be analyzed. We shall focus specifically on distinctive regime characteristics such as the Liberal Democratic Party's (LDP) dominance of the Diet (parliament), the peculiarities of Japan's electoral system, the LDP's grand coalition of interest support, segmented policy configurations, LDP factions, the movement of ex-bureaucrats into the parliament, and LDP interactions with the bureaucracies, especially with MITI. A model of Japanese industrial policy-making will be constructed that explains the curious incongruity between politicized, market-defying policies in the primary sector and nonpoliticized, market-conforming policies in most manufacturing sectors, including high technology.

Continuous LDP Majority Rule

Diet Dominance

The most remarkable feature of Japan's political regime is the LDP's monopolization of majority power in the Diet for almost the entire postwar period. No party in any of the world's large industrial democracies comes close to matching this record. The unprecedented feat has had a profound impact on the structure and processes of Japanese politics. It has given rise to a system in which interest aggregation is determined largely by the political exchange among the LDP, various interest groups in its support coalition, and individual ministries.

The LDP's monopoly of power has to be considered remarkable when viewed against the background of the almost bewildering range of changes that have taken place in Japan's socioeconomic structure: major demographic developments, the evolution of industrial struc-

TABLE 4.1

Distribution of Seats in the
Lower House, 1988

Liberal Democratic Party (LDP)	259
Japan Socialist Party (JSP)	113
Kōmeitō	59
Democratic Socialist Party (DSP)	39
Japan Communist Party (JCP)	27
New Liberal Club	8
Social Democratic League	3
Independents	3

ture, concomitant shifts in occupational distribution, external shocks, and significant institutional change. The LDP has presided over nothing less than a postwar economic metamorphosis. The transformation has had far-reaching ramifications, some of which (like steep increases in real income) have helped to maintain the LDP in power, whereas others (like the contraction of the primary sector) have eroded its traditional base of voter support.

The LDP has managed to hold its Diet majority in spite of the steady decline in its percentage of the popular vote, beginning in the late 1950's and bottoming out two decades later. In 1988, following a major election victory in 1986, the LDP commanded 259 seats out of 511 in the House of Representatives, just over the threshold for a numerical majority. If the New Liberal Club, an LDP offshoot, and Independents are added, the conservative majority comes to 270, a comfortable margin of control. The LDP's 259 seats are more than twice the number of the largest party in the opposition camp, the Japan Socialist Party (JSP), more than three times that of the second-largest opposition party, the Kōmeitō, and nearly ten times that of the Japan Communist Party (JCP). Indeed, the largest LDP faction, with over 60 members in the Lower House, is bigger than any of the opposition parties except the JSP (see Table 4.1).

Slim as the LDP's majority appears, with only 3 seats over the threshold of 256, the numbers alone fail to communicate the true dominance of the ruling conservative party. The New Liberal Club and Independents work closely enough with the LDP almost to be considered de facto members of the ruling party. Moreover, top leaders in the LDP maintain fairly close behind-the-scenes ties with members of the Democratic Socialist Party (DSP) and the Kōmeitō, the third- and fourth-ranking parties, blurring the boundaries between conservative and opposition camps. Even if the LDP should lose its parliamentary majority, as political crystal-ball gazers have been predicting since the late 1950's, it would still be in a position (unless it split into smaller parties) to enter into a formal coalition with one or more of the small opposition parties.

None of the opposition parties is big enough to entertain hopes of unseating the LDP. Owing to deep divisions within the opposition ranks, there is no realistic possibility of a united coalition forming. The differences dividing, say, the Kōmeitō or Democratic Socialists, on the one hand, from the Communists, on the other, are far deeper than those separating the Kōmeitō and Democratic Socialists from the LDP. Although opposition leaders have often talked about the desirability of forming a united front, the talk has never amounted to anything more than wishful thinking. About the best the Kōmeitō, DSP, and JSP have managed to do is agree not to compete against one another in a few election districts.

Party weaknesses and deep schisms within the opposition camp have thus helped the LDP to stay in power. Even voters who are disgruntled with the LDP feel there are no viable alternatives. In their eyes, not one of the opposition parties appears ready to step in and assume the reins of government. In sharp contrast to the LDP, most of the "progressive" parties depend on support primarily from a single, narrowly based, relatively closed interest group. So strong, indeed, are the ties of dependence that the opposition parties tend to be viewed as mere "captives" of their particularistic support groups. To survive in Japan's political marketplace, opposition parties—the JSP, DSP, Kōmeitō, and Japan Communist Party—must maintain the organizational, electoral, and financial backing of their respective support groups—the Sōhyō labor union, Dōmei labor union, Sōka Gakkai (a Buddhist sect), and the Communist party. Yet such dependence comes at a high price. It stands in the way of attempts to broaden the base of popular support beyond the narrow confines of the main support organizations. Nearly all opposition parties are thus stuck with the image of being the captive of some relatively narrow special interests.

The encompassing nature of the LDP's coalition of supporters could hardly pose a greater contrast. The LDP is not only free of domination by a single interest group, but its support coalition consists of groups so diverse that, on certain issues, their interests actually conflict. Being the only party with a broad base of national support places the LDP in a very advantageous position. The LDP has fully exploited the powers that come with incumbency to consolidate the hegemonic structure of Japan's political system.

The Consequences of LDP Hegemony

The LDP's monopolization of parliamentary power has had the effect of strongly reinforcing the status quo. Japan has had a very stable configuration of power since the 1950's. The momentum for change has come primarily from three sources: the rapid growth and maturation of Japan's economy, belated internationalization (including foreign pressures), and societal adjustments triggered by the first two factors. Comparatively little momentum has come from shifts in the distribution of power among the political parties. Opposition parties have had almost no chance of breaking into the inner sanctum of power.

The stability of the status quo is greatly enhanced by Japan's social structure. The overwhelming majority of Japanese—over 90 percent, according to survey data—consider themselves "middle class."[1] Not surprisingly, the proportion expressing satisfaction with their life is also very high, exceeding 85 percent.[2] The sense of satisfaction is no doubt related to the equity of income and wealth distribution. Whatever method of measurement is used—the Gini coefficient (0.316) or the Atkinson index (0.08)—Japan ranks at or near the top of all OECD countries in terms of equitable income distribution.[3] Hence, on the basis of structural and attitudinal indicators, Japanese society appears to be among the world's wealthiest and most stable. It is no accident that Japanese politics has also been so stable.

Equity in income distribution suggests that the beneficiaries of the existing status quo extend beyond the circle of the LDP's ruling coalition. Even "excluded" groups, like organized labor, that support the JSP and DSP have realized dramatic gains in real income during the decades of conservative party rule. Indeed, the rate of increase in real, inflation-adjusted wages in manufacturing has been faster in Japan than in most other OECD countries. At the same time, wage differentials by sector, occupation, and size of firm have also tended to narrow.[4] Even though mainstream interests, like big business and finance, have reaped a disproportionate share, the non-mainstream groups, like small and medium-sized businesses and retail distributors, have not been totally shut off from the benefits of economic growth. This may be the secret to Japan's political stability. Because the benefits of rapid economic growth have been spread equitably, including even groups that stand outside the nimbus of the conservative coalition, a large proportion of the public has come to hold at least some stake in upholding the fundamental structure of the system.

The system of single-party hegemony has also spared Japan from the policy oscillations that occur when one party or a majority coalition replaces another, as in the United States. Under the LDP, Japan has never wavered in the pursuit of policies aimed at fostering growth, equity, improvements in the quality of life, and economic security. As a result, Japanese corporate executives—unlike their U.S., French, and British counterparts—operate in a predictable business environment. They have not had to worry about abrupt shifts in government policies. Japan's electronics industry, as pointed out in Chapter 2, has not had to cope with varying taxes on capital gains, which, in the United States, rose as high as 40 percent before dropping to 18 percent. For medium- and long-term corporate planning, the value of a steady, predictable public policy orientation can hardly be exaggerated.

The LDP's postwar dominance has also enabled it to develop a mutually satisfactory, routinized relationship with MITI. On most issues related to industrial policy, the LDP allows MITI extraordinary scope for autonomous action. Few if any other ministries inside or outside Japan enjoy the same degree of freedom. Most other ministries in Japan—Construction, Transportation, Health and Welfare, Post and Telecommunications—have to operate within more circumscribed administrative and political boundaries.

MITI's greater leeway can be attributed to a number of factors: (1) the capacity of MITI and industry to reach consensus; (2) the wide scope of MITI's authority over the manufacturing sectors, which protects it against any "salami slice" tactics on the part of interest groups and the LDP; (3) MITI's emphasis on promotional rather than regulatory policies (in contrast to, say, the ministries of Post and Telecommunications and Health and Welfare); (4) MITI's minimal reliance on formal legislation that must be passed through the Diet; (5) MITI's capacity to use informal persuasion and administrative discretion instead; (6) MITI's relatively small budget for public procurements (unlike Construction, Transportation, Agriculture, and Defense), which limits opportunities for pork-barrel politicking; and (7) the electoral importance of labor-intensive groups, like farmers, in the LDP's grand coalition. Other factors will be discussed later in connection with a model of industrial policy-making in Japan.

Of the seven factors, special weight should be placed on the first and seventh. So long as MITI and industry can work together to achieve common goals that advance private and public interests, there is little reason for either side to involve the LDP. It is only when industry feels it cannot work out mutually acceptable industrial policies with MITI, or when MITI must secure Diet approval for budgetary allocations and formal legislation, that incentives arise to turn to the LDP for political assistance. Even then, the temptation is usually tempered by the calculus of structural interdependence and long-term, reciprocal obligation. Private companies realize that they will be called upon to make monetary contributions if they ask for political assistance; likewise, MITI is aware of its vulnerability to political pressures when it relies too heavily on the LDP.

LDP Inclusivity

Japan's one-party-dominant system rests on the LDP's broad base of electoral support, covering a diverse cross-section of society—from farmers to big business executives and small and medium-sized entrepreneurs. The LDP even picks up votes from rank-and-file members of organizations like Dōmei, which is formally aligned with the DSP. The electoral base of the LDP's support coalition thus spans nearly all segments of society in at least some measure, bridging what are often deep cleavages in other countries: between young and old, labor- and capital-intensive industries, primary and manufacturing sectors, rural and urban, big and small businesses, self-employed professionals and housewives, management and labor. Although the proportion of those represented from each category varies, of course, the LDP's is truly a grand coalition, if the concept is defined in terms of the scope and diversity of interests represented.[5]

The inclusivity of the LDP's ruling coalition parallels, in certain respects, bureaucratic inclusivity. Virtually all interest groups in Japan fall under the jurisdiction of a government bureaucracy.[6] The parallelism simplifies interactions between the legislative and administrative branches. Instead of having to divide their time evenly between conservative and opposition camps, Japanese bureaucrats can concentrate on working with key LDP leaders, who head LDP factions or policy committees or who belong to various informal caucuses (*giin renmei*) and policy support groups (*zoku*). Because ad hoc LDP policy groups cover the full range of economic activities, the inclusive, hegemonic structure functions fairly smoothly to aggregate interests with respect to bread-and-butter issues.

Political and administrative inclusivity is made possible by the homogeneity of Japanese society, especially the comparative absence of deep cleavages. There are few permanent schisms based on class, ideology, religion, region, race, history, ethnicity, or language. The religious schism between clericalism and anticlericalism historically observed in France and Italy, for example, has no counterpart in Japan.[7] The traditional antagonism between labor and management, perhaps the most divisive political cleavage in the West, is muted in Japan by the company-based nature of labor union organization and the consensual style of company management. The LDP's ability to hold together and even extend a diverse coalition of interests has also been greatly enhanced by the positive political consequences of sustained economic growth, especially the steep increases in real income for the Japanese people, combined with the equity of income distribution.[8]

Overrepresentation of Rural and Semiurban Districts

Japan's system of electoral districting has created a situation in which voters in rural and semiurban constituencies have had a greater voice in deciding who gets elected than their counterparts in metropolitan districts. Although eligible voters in the agricultural sector made up around only around 20 percent of the national electorate in 1976, the rural and semiurban districts in Japan decided about 30 percent of the seats in the Lower House of parliament.[9]

The overrepresentation of rural voters is seen in the big discrepancies in the number of votes it takes to win a seat in the Lower House. In the 1982 election, for example, a vote cast in the 5th District of Hyogo Prefecture carried more than three times the weight of a vote cast in the 4th District of Chiba Prefecture; that is, a candidate from Hyogo's 5th District won a seat with only 81,375 votes, compared to the 321,351 it took to win one in the 4th District of Chiba.[10] Thus, Japan's system of election districting has allowed agriculture to be disproportionately represented relative to its percentage of the population. The egregious imbalance, which calls into question the constitutionality of election districting, has continued with only cosmetic adjustments (adding some seats to the metropolitan and urban districts). The tyranny of the status quo is reflected here in the fact that the two largest parties, the LDP and JSP, have resisted plans for radical reapportionment because both benefit from the current imbalance.

The LDP (as well as the JSP) has had its strongest base of support in rural and semiurban districts, with the small opposition parties—the JCP, Komeito, and DSP—faring best in the large metropolitan districts (see Figs. 4.1 and 4.2). If one vote carried the same weight in metro-

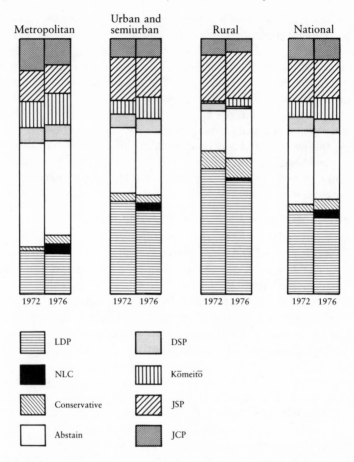

Fig. 4.1. Shifts in party support by type of electoral district in the 1970's: Lower House. Source: Ishikawa Masumi, *Sengo seiji kōzōshi* (A history of the postwar structure of politics) (Tokyo: Nihon Hyōronsha, 1978), p. 112.

politan precincts as in the rural districts, the small opposition parties would be in a position to claim significantly more seats in the Diet without any change in the percentage of votes they receive. As it is, only 25 districts out of 130 can be classified as "purely" metropolitan. Masumi Ishikawa points out that although the population concentrated in these metropolitan precincts—in Tokyo, Yokohama, Osaka, Nagoya, Kyoto, and Kobe—represents more than 40 percent of the national total, it elects only 100 representatives out of 511 in the Diet, or less than 20 percent.[11]

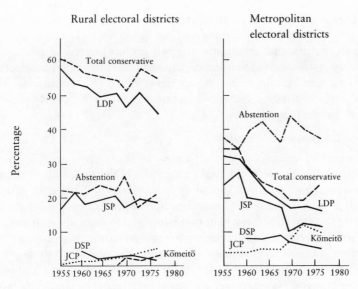

Fig. 4.2. Trends in party support, rural and metropolitan districts, 1955–1980. Source: same as Fig. 4.1.

Through shrewd campaign strategy, the LDP can usually capture a stable share of seats from the metropolitan districts. This is due to the high percentage of "dead" or "wasted" votes (that is, votes cast for candidates who fail to be elected) in the metropolitan precincts. In contrast, a relatively low percentage of "dead" votes (about 20 percent of the total) are cast in rural districts. By minimizing its number of "dead" votes, the LDP is usually able to win around one-third of the seats from the 25 metropolitan districts.[12]

Hence, the power of the agricultural sector far exceeds its demographic size, to say nothing of its economic output. According to Kenzō Hemmi, "the LDP has been able to remain in power for so long because, and only because, it has enjoyed support in rural districts rather than in the large cities."[13] The disproportionate allocation of seats is the most immediate reason that the LDP has managed to hold on to its parliamentary majority in spite of the fact that its share of the popular vote has fallen below 50 percent.

Halting the Slide

For two decades, from the late 1950's to the late 1970's, the LDP found its percentage of votes falling steadily from above 50 percent to well below. To arrest its slide at the polls, the LDP adopted in the late 1960's a two-pronged strategy of revitalization: (1) consolidating clientelistic ties with traditional support groups located largely in the primary sector and (2) wooing disaffected voters through the selective correction of certain public policy deficiencies. In combination, the two-pronged strategy succeeded in bringing the LDP's steady slide to a halt by the late 1970's.[14]

Consolidation of Clientelism

As part of the first course of action, to shore up its sagging competitiveness in the political marketplace, the LDP stepped up its efforts to reaffirm ties with traditional support groups.[15] The main instrument used was the expansion of public expenditures for politically targeted sectors, an instrument that happened to coincide with macroeconomic circumstances requiring greater government spending.

In the wake of the first oil crisis in 1973–74, the sharp drop-off in private investment placed Japan in the then unusual situation of having an excess of savings over business investment demand. The large supply of excess savings forced the Japanese government to pursue more vigorously an interventionist role in order to avert the very real danger of a severe recession.[16] The LDP government adopted an orthodox Keynesian policy of demand stimulation, featuring vigorous government spending in areas of obvious need, such as public works construction and the development of social overhead infrastructure. Japan's macroeconomic needs happened to dovetail, therefore, with the perceived need to implement an aggressive program of clientelistic targeting.

By relying heavily on deficit spending, the LDP government managed to absorb a significant portion of the country's excess savings and to pump vast amounts into the economy. Public expenditures rose from ¥8 trillion in 1970 to ¥20 trillion in 1975 and ¥50 trillion in 1984. This was more than a sixfold increase over only a fifteen-year span. The targeted sectors of pressing political concern, like construction, local finance, agriculture, and small and medium-sized enterprise, all shared in the massive flow of public spending.

Allocations for public works—which benefit traditional LDP support groups like local developers and the equipment, housing, construction, lumber, and cement industries—rose from ¥1.4 trillion in 1970 to ¥7.2 trillion in 1982.[17] Funds for local finance rose from ¥1.7 trillion in 1970 to ¥9 trillion in 1982. And public expenditures for foodstuffs control increased from ¥488 billion in 1970 to ¥1

trillion in 1980, declining to ¥917 billion by 1983. The dramatic
expansion of public expenditures, targeted especially to local interest
groups in the LDP's coalition, had the effect of rewarding the conservative party's traditional core of faithful followers, groups hard hit by
the slowdown in growth rates or by the loss of international comparative advantage.

The LDP also actively sought to consolidate and broaden its circle
of supporters by wooing other sectors of society that were either outside its traditional coalition or incompletely incorporated. In particular, the LDP targeted interest groups that, for business reasons, seemed
likely to swing toward the opposition camp.[18] One prime target was
small- and medium-scale business (*chūshō kigyō*), a vital sector of the
economy and a significant factor at the polls, with nearly 18 million
members, or twice the membership of Nōkyō, Japan's powerful agricultural cooperative. In the late 1960's, the LDP began to worry about
the political drift of small and medium-sized enterprise as the JCP
infiltrated this sector through the organization of the People's Chamber of Commerce (Minshū shōkōkai, or Minshō). The hardships under which small businesses were forced to operate made this sector a
potentially fertile ground for voter recruitment by the JCP.

To compensate for the handicaps under which small-scale businesses operate, and more important, to ensure that it would not lose
this large and politically important sector to the communists, the LDP
took vigorous steps to assist small and medium-sized enterprise in
order to bring the sector more fully into the fold.[19] The LDP granted
lenient tax provisions, turned a blind eye to large-scale tax evasion,
and made special loans available without collateral at 7 percent interest. The sum of loans extended to small businesses skyrocketed from
¥30 billion in 1973 to ¥510 billion in 1980, an incredible seventeen-
fold increase. At the same time, allocations for small businesses from
the General Accounts Budget jumped from ¥50 billion in 1970 to
¥239 billion in 1983. The dramatic rise in the combination of tax
breaks and tolerance of significant tax evasion, soft loans, and budgetary allocations conveys some sense of the tremendous importance
LDP leaders attached to maintaining the support of the small and
medium-sized enterprise sector.

With the LDP paying off faithful support groups through the provision of the usual rewards associated with pork barreling—subsidies,
public works contracts, procurements, allocations from the General
Accounts Budget, tax breaks, protection against foreign competition,
administrative guidance (for example, concerning the stabilization of

prices), and favorable legislation—LDP support groups in the labor-intensive sectors—farmers, fishermen, local construction firms, real estate interests, distributors, small retailers, and others—reaped the benefits of a continual redistribution of income as Japan's industrial economy expanded. In consequence, the gap between city and countryside has never developed into a threat to political or social stability.

On the negative side, the pork-barrel attention paid to the labor-intensive sectors and the LDP's dependence on financial contributions from big business have led to a variety of costs: abuses of power, corruption, and the emergence of more than a few enclaves of economic inefficiency in the primary and service sectors. To the consumer public have fallen the costs of political clientelism: exorbitant food prices, incredible land prices, inadequate housing, and a lack of international competitiveness in various areas of the primary and service sectors.

With multiple enclaves of inefficiency, it is surprising that the healthier sectors of the economy have not chafed under the weight of transferred costs (such as international trade frictions). The inefficiency of pork-barrel politics is offset by the relative lightness of burdens that other countries have had to bear, such as heavy defense expenditures. There is some slack within Japan's political economy, in other words, to absorb the nontrivial costs of catering to the parochial interests of the LDP's traditional, clientelistic support groups.

The Japanese public appears willing to accept the economic costs of political inclusivity. There is widespread sympathy, for example, for the farmers' declining competitiveness and structural predicament, since agriculture has been the mainstay of the Japanese economy for centuries and the majority of Japanese still have direct family or kinship (only one generation removed) ties with farming. Concern about food self-sufficiency also disposes the Japanese to favor domestic food production, in spite of the availability of less expensive food from abroad.[20] Add to this the public's strong sense that all groups deserve equitable treatment, and the willingness to spread or collectivize the costs of inefficiency can be understood.

For decades the service sectors, however inefficient, absorbed excess labor and, in the case of local mom-and-pop retail stores, provided customers with conveniences that seemed to offset the aggregate costs. More than half of all retail outlets in Japan employ only one or two people; nearly 40 percent of the wholesale outlets employ four or fewer. Given a relatively early retirement age of 55 and the restricted mobility of Japanese housewives, such retail networks have served

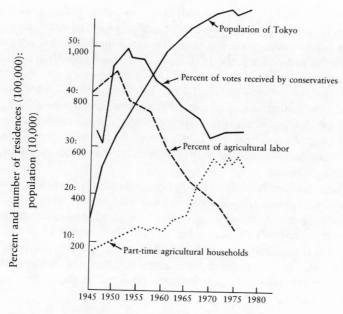

Fig. 4.3. Party support, population, and the agricultural sector, 1945–1980. On the vertical axis, the first number indicates either the percent or the number of households expressed in 100,000's; the second number indicates the population expressed in 10,000's. Source: same as Fig. 4.1, p. 184.

social and labor market functions that the LDP has found to be politically useful.

From the point of view of market economics, government subsidies and side payments to economically inefficient sectors may seem unjustifiable, since finite capital resources can be put to use in more productive ways. However, from a social and political point of view, LDP leaders can justify the expenditures as a kind of social welfare investment, a collectivization of economic costs that serves several purposes. It redistributes income to the remote regions of Japan, which have not shared in the expansion of industrial output. It helps small businesses cope with the disadvantages of being small and operating outside the mainstream of business. It fulfills the government's responsibility to promote economic equity. It gives small businesses a better chance of riding out the ups and downs of business cycles. And it stabilizes the status quo and consolidates the structure of political hegemony.

An irony of Japan's electoral system is that the primary sector's support for the LDP permitted MITI to implement policies that stimulated industrial growth—which in turn led to the steady contraction of the primary sector (see Fig. 4.3). Although the fruits of double-digit growth helped to keep the LDP in power, rapid growth altered and undermined the foundations of the LDP's traditional coalition. Structural transformation—the shift from smokestack to high-technology sectors, the redistribution of capital and labor, urbanization, and concomitant changes in attitudes—has eroded the foundations of single-party hegemony, causing the LDP's share of votes to drop and forcing it to make policy adaptations that have stretched the boundaries of its political coalition.

Electoral support for the LDP has declined steadily over the postwar period, paralleling the shrinkage of the agricultural labor force and moving in just the opposite direction of the trend toward urbanization as seen in Tokyo's demographic growth. It was not until the late 1970's that the LDP's decline finally bottomed out. This indicates that the very successes of rapid growth served to transform the foundations of LDP support, most conspicuously reflected in the decline of the primary sector.

Corrective Public Policy

The second part of the LDP's two-pronged strategy of revitalization was to tackle the three most pressing and politically embarrassing problems of public policy facing Japan during the late 1960's and early 1970's: environmental pollution, social welfare, and inflation. By correcting these public policy problems, which left the LDP vulnerable to attack by opposition parties, the ruling conservative party hoped to placate enough disaffected voters to stem the decline of its popularity.[21]

Public outcry concerning environmental pollution, which reached a crescendo between 1968 and 1972, died down after the government tightened regulatory requirements and brought levels of carbon monoxide, sulfur, and nitrogen oxide under control. By 1975, the pollution issue had all but disappeared from the agenda of serious societal problems in urgent need of public policy correction. This denied the opposition camp one of its easiest targets of attack.

The LDP also managed to preempt opposition camp criticisms of the inadequacy of social welfare programs, another thorn in the conservative party's side. The remedy, as in the case of consolidating ties with traditional support groups, consisted of pouring massive amounts of money into welfare. As seen in Figure 4.4, public expen-

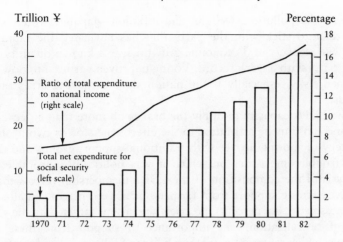

Fig. 4.4. Total Japanese expenditures for social security, 1970–1982. Source: Japan, Prime Minister's Office, *Statistical Handbook of Japan* (Tokyo, 1984), p. 121.

ditures for social security increased ninefold from 1970 to 1982, rising from around ¥4 trillion to nearly ¥36 trillion. In terms of the share of national income, this represented a net jump from less than 2 percent to more than 16 percent, or an eightfold increase. Because pollution and social welfare stood out as the LDP's most embarrassing public policy failures, the LDP's capacity to remove them from the glare of public scrutiny and the calculus of electoral accountability helped to stop its slide in popularity.

The Japanese government similarly succeeded in bringing runaway inflation under control. Loose monetary policy prior to 1973 exacerbated the inflationary impact of the first oil crisis; but stringent controls on money supply prior to 1979, the year of the second oil crisis, cushioned that crisis's effects. The turnaround can be explained in part by historical happenstance—bad timing and circumstances prior to the first oil shock and fortunate timing and circumstances prior to the second. But the different reaction to the second oil crisis demonstrated that the Japanese government could learn from its mistakes. Since the mid-1970's, Japan has had a comparatively low rate of inflation owing largely to sound macroeconomic management and the appreciation of the yen.[22] Again, sound public policy helped restore public confidence in the LDP.

Impending national crisis, like the two oil shocks during the 1970's, seems to bring out the basic conservatism of the Japanese people. Unless the LDP demonstrates that it is incapable of handling the tough issues—like pollution, welfare, and inflation—Japanese voters seem to prefer to stick with the party that has navigated the rapids of postwar recovery and economic growth. It is a known quantity, the only party with a track record. Voting unproven parties into power is deemed risky, especially if the nation were to have a crisis on its hands.[23]

The LDP's capacity to apply the brakes on more than a decade of decline can thus be attributed to the effectiveness of its two-pronged strategy of shoring up ties with traditional support groups and tackling troublesome issues on which it was politically vulnerable. An analysis of the composition of the LDP's membership reveals that it received strong support from farmers (27.1 percent) and small-scale businessmen (28.4 percent), precisely the groups targeted by the strategy of revitalization. The importance of support from these two groups, which represent more than half the LDP's membership of 2.57 million in 1988, is illustrated by the fact that they account for only around 20 percent of the national electorate. The LDP's capacity to command and hold their allegiance is noteworthy, given the fact that both groups have been beset by serious structural problems—small-scale business by size disadvantages and sensitivity to business cycles, agriculture by the loss of comparative advantage and sustained pressure from foreign producers. It is precisely because of the severity of these problems that these two groups gravitated to the LDP, the only party capable of using the full instruments of incumbency to respond to their needs.

In addition to consolidating clientelistic ties with traditional support groups, therefore, the LDP's strategy of investing in new public policy initiatives also paid handsome dividends in terms of stretching the boundaries of its support coalition. Public opinion polls indicate that the LDP has made noticeable inroads into groups that used to be distant or hostile: metropolitan residents, workers with only a junior high or a high school education, young adults, unskilled workers, the self-employed, and housewives. Such groups had remained outside the LDP fold, feeling dissatisfied with government policies and the patterns of interest aggregation.[24] As Yasusuke Murakami writes, "The conservative support from the groups which had previously been against the LDP such as the urban, the young, and the [self-] employed people significantly increased from 1973 to 1978, and this new support tended to be passive in nature."[25] The LDP has thus broadened its base of electoral support, but the voters' explicit linking of expected benefits to election support for the LDP has made the party's majority less stable and less predictable than before.

Among the new supporters, party identity tends to be weaker and election turnout lower than those of traditional supporters like farmers. Behind the LDP's resilience, therefore, is a combination of active, hard-core, traditional party loyalists from the rural and semiurban areas and a smaller group of less clearly identifiable, less committed, more passive "floating" supporters from the urban and metropolitan districts. The support of both groups can be traced, at least in part, to the effectiveness of the LDP's catch-all interest-oriented strategy.

The targeting strategy adopted by the LDP appears to conform to the logic of exploiting the powers of incumbency to the fullest by distributing material benefits widely while at the same time focusing on key support groups.[26] Murakami attributes the LDP's capacity to ride out both the turbulence of short-term crises and the speed of long-term structural transformation to its ability to convert itself from a tradition-oriented to an interest-oriented, catchall party.[27] The LDP's responsiveness to the tangible interests and concerns of Japan's new "middle mass" has expanded its political inclusivity, broadened its base of voter support, and extended the life expectancy of its hegemonic power. The LDP's continued dominance, in turn, has continued to have the positive effects on industrial policy discussed earlier.

Political Exchange

The LDP's successful strategy of revitalization provides an insight into the nature of the ruling party's relationship with key interest groups in its support coalition. The dynamics of LDP interactions with its interest coalition and with the bureaucracy, in turn, supply the essential materials for constructing a typology of the Japanese policy-making processes, called here segmented political configurations. The typology explains why there are striking variations across sectors in the degree to which Japanese industrial policy is politicized. It helps us understand why industrial policies for the manufacturing and high-technology sectors under MITI have been relatively free from political interference, while those for agriculture and construction tend to be dictated by political considerations. The taxonomy is based on the fundamental concept that political transactions, involving the exchange of political goods and services, serve as the driving force behind the nexus of ties binding the LDP to the various interest groups in its support coalition. Such transactions take place under a single-party-dominant political system, within the framework of jurisdictional boundaries marking off the individual ministries.

Over decades of single-party rule, political exchanges have become relatively routinized. The inability of opposition parties to pose a realistic alternative to single-party rule has been of overriding significance. It has meant that most interest groups in the LDP's grand coalition have had no choice but to depend on the LDP, operating within an institutional framework of policy segmentation. Out of the complex interplay among groups in the LDP's grand coalition, the LDP itself, and the bureaucracies has emerged the typology of political exchange. The LDP's grand coalition can be divided into the following four categories, based on the nature of the political goods and services exchanged.

Clientelistic Votes. The LDP receives votes from clientelistic support groups in exchange for favorable legislation, subsidies, generous tax treatment, and other promotional policies. Private interests that take part in the exchange include traditional LDP support groups—agriculture, small-scale enterprises, health professions, heads of local postal services, local real estate interests, and so forth. The ministries involved include Agriculture, Fisheries and Forestry, Agency for Small and Medium-sized Enterprise (MITI), Post and Telecommunications, and Health and Welfare.

Reciprocal (Pork-Barrel) Patronage. The spoils of public expenditures (public works, procurements, subsidies) are recycled back to the LDP in the form of financial contributions. The relationship is one of reciprocal patronage. Involved interest groups include a variety of traditional LDP supporters—local interests, construction, transportation, defense industries. Recycling also takes place in the regulatory arena, with regulated industries contributing to the LDP in return for acceptable regulatory policies (telecommunications, electrical power utilities). The bureaucracies involved include Construction, Transportation, Post and Telecommunications, Local Autonomy, and Defense.

Untied Financial Support. Various interest groups—big business, banking and financial institutions—give the LDP financial support, untied to specific expenditures or public policy favors. It can be understood as a general exchange, one based on the broad benefits that the business community receives from having a pro-business party in power. Big businesses' willingness to contribute what are usually large sums—in spite of the public goods nature of the contribution—can perhaps be understood as payment of a kind of insurance premium by those most capable of paying. The bureaucracies under whose jurisdiction these donors fall are MITI and MOF and various divisions or bureaus in other ministries.

Generalized Voter Support. As the size of groups in the first category shrinks with economic maturation, the LDP has had to secure support from a much broader and more diffuse cross-section of voters: white-collar salarymen, housewives, young residents in metropolitan areas, the self-employed. These groups tend not to be politically well organized, to identify themselves only weakly (and conditionally) with the LDP, and to vote on the basis of general policy concerns (like the overall health of the economy, welfare, and pollution). The LDP tries to appeal to them through the crafting of public policies aimed at improving the overall quality of life. Given the diffuseness of the groups and their concerns, no single set of ministries can be identified as principally in charge.

Crude though the typology is, it provides a much clearer, more concrete model of the Japanese policy-making processes than any of the standard theories available in the social science literature: elitism, liberal pluralism, neocorporatism, or statism. None of these four theories explains the hows and whys of the policy-making processes in Japan. It is hard to reconcile the theory of elites with abundant empirical evidence of pluralistic competition; elite theory cannot account for the strong Japanese preference for consensual modes of decision making, the overriding emphasis placed on the well-being of the collectivity, or the political elite's quest to maximize the collective good and safeguard public interests.[28] Similarly, liberal pluralism has trouble explaining such features of Japanese policy-making as extensive and informal elite networks, blurred lines of demarcation between the public and private sectors, the functional importance of the intermediate zone between the state and private enterprise, and evidence of political cartelization in certain policy domains.[29]

Neocorporatism cannot account for certain anomalies in Japan, especially the comparative weakness of organized labor unions, their exclusion from the LDP's ruling coalition, and the benefits in which they nonetheless share.[30] Nor are peak associations as powerful in Japan as they are in parts of Europe. Keidanren, for example, the most powerful peak association in Japan, commonly portrayed as the citadel of big business, exercises much less power than is often inferred from the size and prestige of its corporate members.[31] Statist theories are unable to explain the strength of the private sector in Japan, the numerous instances of private sector recalcitrance in the face of administrative guidance, and the steady erosion of bureaucratic power.[32] A closer approximation to the realities of policy-making in Japan requires a model based on the dynamics of political interactions among LDP, its interest coalition, and the bureaucracies.

Type 1: Clientelistic Linkages

From the standpoint of the LDP, Type 1 is probably the most important form of political exchange, since the LDP's standing as the majority party hinges on solid support from traditional groups capable of delivering votes—farmers, fishermen, small- and medium-scale businessmen, heads of local postal services, the medical professions, and so forth. In nearly all Type 1 cases, a fairly stable structure of relations among the LDP, interest groups, and bureaucracies has taken shape over decades of routinized interaction. Type 1 interest groups—particularly agriculture, the medical professions, small-scale businesses, war veterans, and heads of local postal services—have organized themselves effectively in order to maximize their bargaining power vis-à-vis the LDP and bureaucracies. They all have powerful LDP *zoku* (informal policy support groups) behind them to lobby on their behalf not only in the Diet but also with the relevant ministries. Since the ministries in charge of Type 1 industries are fairly narrowly focused, Type 1 interest groups tend to have more leverage than interest groups in Types 2–4. Both the LDP and the ministries try to be responsive to their needs.

From the LDP's point of view, the value of the political exchanges that take place in Type 1 sectors is enormous, given Japan's electoral malapportionment and the LDP's dependence on clientelistic groups for its parliamentary majority. What the LDP stands to gain—crucial votes—is of such value that it must come up with political goods and services of comparable value to reward its faithful supporters. The LDP's package of political payments is quite generous when measured in terms of the economic inefficiencies that are tolerated. For agriculture, the payments include protection from foreign imports, price controls, and large subsidies. For small-scale businessmen, as pointed out earlier, the package includes special low-interest loans, generous tax treatment, and a variety of consulting services. For the medical profession, the exchange includes generous health and welfare insurance coverage, preferential tax treatment, and a strong voice on all policy issues related to medicine and health care. The political bargains struck in Type 1 policy domains often come at a high cost to consumer interests and overall economic efficiency.

Type 2: Reciprocal Patronage

Closely related to the Type 1 exchange is the Type 2 form of transaction, which involves the recycling of political patronage, or what Michisada Hirose has aptly called the "political profit-sharing system" (*rieki haibun shisutemu*).[33] Industries falling into this category are construction, housing, highways and roads, real estate, regional and local development, defense contracts, transportation, tobacco, the electrical power utilities, and telecommunications. Most of these industries are capable of mobilizing votes in local constituencies. Construction companies, for example, are sometimes central political actors in rural and semiurban districts. For LDP Diet members from certain election districts, the organizational and financial support from construction companies may determine whether they win or lose.[34] In exchange for generous financial and organizational backing, construction companies and local developers receive LDP assistance in securing government authorization for the use of land, licensing rights, public works contracts, and a variety of other services. The relationship is one of reciprocal patronage.

What the LDP offers support groups in Type 2 transactions can be divided into two subtypes: Type 2a consists of procurement and public works contracts, budgetary allocations, and subsidies; Type 2b involves acceptable regulatory controls, facilitation of administrative procedures, and legislative support. The Ministry of Construction and, to a lesser extent, the Ministry of Transportation are the most heavily involved in Type 2a transactions—procurements and public works; the bureaucracies most commonly associated with Type 2b activities—regulatory oversight—include Health and Welfare, Post and Telecommunications, and the Agency of Natural Resources and Energy (affiliated with MITI). All five have extensive dealings with the LDP, but Type 2a industries, bureaucracies, and LDP caucuses tend to be more politicized than those of Type 2b. The Ministry of Construction may be the most politicized of all bureaucracies in Japan.

The distinction between the two types of reciprocal patronage should be applied to private companies that depend on public works and government licensing, like construction (Type 2a), as well as regulated government monopolies, like the electrical power utilities (Type 2b). Although both types contribute funds to the LDP, they differ in the nature of their relations to the LDP bureaucracies and in the extent to which they are politicized. The electrical power utilities and telecommunications are more powerful, less politicized, and more independent. Indeed, in some respects, the LDP depends as much on these regulated monopolies as these regulated industries do on the LDP. Apart from their political contributions, the electrical power utilities have an enormous impact on the economic well-being of local areas, not only because of the generation of electrical power but also because of the magnitude of their capital investments, which annually account for nearly 10 percent of the country's total. Unlike their non-partisan brethren in the United States, the electrical power utilities in Japan actively support the conservative party and can be considered part of its inclusionary coalition of interests.

In the case of the electrical power utilities, unlike that of most Type 2a companies, ties with the LDP are intermediated by MITI, the regulatory agency under whose jurisdiction utility companies fall. For reasons that have already been discussed, MITI is more successful at fending off political interference than most other bureaucracies. Since MITI sets utility rates and regulates the public utility companies, and since the industry is itself strong, the LDP has not been able to dominate the regulated monopolies in the same way that it controls construction, local development, and other Type 2a industries. Nonetheless, Type 2b industries still fall into the broader pattern of reciprocal patronage because they funnel donations and voter support to the LDP in exchange for favorable regulatory policies.

The Type 2 ministries tend to be dominated by domestic interests and concerns, given the fact that few industries under their jurisdiction are dependent on overseas markets. Only shipbuilding, construction, and telecommunications sell a substantial portion of what they produce to overseas customers. Most of the other industries—home construction, pharmaceuticals, and defense procurements (Type 2a) and telecommunications, electrical power utilities, and transportation (Type 2b)—rely primarily on domestic markets. In Type 2 bureaucracies (except the Agency of Natural Resources and Energy), the voice of internationally minded officials is conspicuously weak. Bureaucratic officials are concerned primarily with protecting the well-being of domestic producers. They tend not to be concerned about, or indeed conscious of, the international ramifications of domestic industrial policy.

As in the case of Type 1 political configurations, the political exchange—reciprocal patronage—often takes place at the public's expense. Such transactions always run the risk of corruption, abuses of power, political cartelization, and economic inefficiency. However, Type 2 differs from Type 1 by virtue of the fact that most Type 2 industries have not yet lost international comparative advantage (unlike agriculture and the primary sector). This means there is less need to resort to import protectionism (though that exists to some extent) and therefore less loss of economic efficiency.

Type 3: Generalized Support

Type 3 power configurations differ from Type 1 in that they involve neither the primary sector nor the mobilization of votes. They differ from Type 2 configurations in that the political exchange between the LDP and industry is not tied to specific commodities—procurements, public works, or regulatory policies. Individual industries receive favorable treatment, to be sure, as part of the LDP's pro-business orientation. But the political exchange is less particularistic and more general in nature, involving the LDP's provision of a collective good— a stable, pro-business environment—in return for the constant infusion of funds, which gives the LDP the wherewithal to stay in power. Big business pays a disproportionate share of the cost of maintaining the LDP in power; but it reaps a disproportionate share of the benefits and it can draw on more resources than most other interest groups in the LDP's inclusionary coalition. It is therefore willing to pay a lot to keep a party so favorably disposed to its interests in office.

Big business can funnel money to the LDP through a maze of channels: from individual companies to individual LDP representatives, from individual business leaders to particular faction bosses, from fund-raising functions (such as receptions) to LDP leaders on the rise, from individual corporations through industrial associations to LDP members of informal policy caucuses, and from keiretsu groups to the party and to prominent leaders, as well as through various "laundered" routes. Perhaps the best-known channel is that between Keidanren (Federation of Economic Organizations) and the LDP. Every year, following careful consultations, Keidanren sets a target for political funds from the business community. It then proceeds to designate how much each industry will contribute on the basis of a complex formula for equitable allocation.[35]

Obviously, the relationship between Type 3 big businesses and the LDP is not uniform across all sectors. Some industries, like steel, are closer to the LDP and politically more influential than others, like the precision equipment manufacturers. As a rule, the more dependent a Type 3 industry is on the bureaucracy (MITI and MOF), the more "strategic" its economic role, the thicker its nexus of linkages with government, and the greater its political leverage.[36] However, the generalization must be modified by the specific characteristics of each industry and the stage of development in its industrial life cycle. High-priority industries at an early stage of growth, like the information industries, tend to keep a greater distance from politics than might be predicted from the standpoint of their economic importance.

Type 3 industries generally maintain a greater distance from the LDP than Types 1 and 2. For example, although the banking industry contributes generously to LDP coffers, it tries to stay clear of the parochial entanglements of electoral politics because banking is considered an enterprise of public responsibility and trust, somewhat akin to the civil service. Or consider the steel industry, politically among the most influential in the Type 3 configuration: it tries to stand above politics and take a broad view of public policy, one that gives proper priority to long-term national interests. Even industries caught up in their own narrow interests that find themselves in need of LDP lobbying vis-à-vis MITI tend to be reluctant to move into too close an embrace with the LDP. By going to the LDP for help, these industries leave themselves open to requests for greater donations.

From the standpoint of the LDP, the arm's-length relationship is perfectly acceptable. Distance does nothing to diminish the value of the Type 3 political transactions. So long as big business continues to supply funds to cover the enormous expanses of electoral politics, the LDP is content. In fact, an arm's-length relationship may be preferable to one of close adhesion. The LDP does not have unlimited time and energy to look after all the interest groups in its grand coalition. By keeping an arm's-length relationship with Type 3 industries, it can concentrate on servicing the clientelistic needs of Type 1 and 2 interest groups, which expect to be accorded special attention. The generalized nature of the Type 3 political exchange thus permits a mutually beneficial relationship to exist at a distance.

This implies that although big business pays more than its share of political dues as a member of the LPD's grand coalition, it is not locked in the same kind of symbiosis with the LDP as agriculture or the same kind of clientelistic relationship as construction. The fact that most Type 3 industries fall into MITI's or MOF's bailiwick reinforces their autonomy vis-à-vis the LDP, since MITI and MOF are the most powerful and least politicized bureaucracies. The LDP cannot control MITI or MOF, or the industries under their jurisdiction, in the same way that it can influence, say, the Ministry of Construction.

There are, of course, units within both MITI and MOF that are more politicized than others. Pockets of politicization in MITI include the Agency for Small and Medium-Sized Enterprise and, to a lesser extent, the Agency of Energy and Natural Resources; in MOF, they include the Japan Tobacco and Salt Public Corporation and, to a lesser extent, the Budget Bureau. Notwithstanding these somewhat isolated pockets of political infiltration, the industries under the two ministries are freer of political interference than those of Types 1 and 2, owing largely to the power of the two ministries.

The essential distinction between Types 2b and 3, on the one hand, and Types 1 and 2a, on the other, can be summed up very simply: whereas the *LDP–interest group nexus* is the critical linkage for Types 1 and 2a, the *bureaucracy-industry relationship* is the central linkage for Types 2b and 3. The structure of Type 2b-3 configurations is thus more conducive to development of a "rational" industrial policy based on long-term technical criteria rather than short-term political imperatives. The distinction is graphically captured in the contrast between Japanese industrial policy for high technology and that for the silkwork industry, national railways, construction, and agriculture.

Type 4: Public Policy Feedback

Type 4 transactions, involving the exchange of targeted public policies for impersonal, nonclientelistic votes, has grown in importance as the size of Type 1 support groups (especially agriculture) has shrunk. The swelling number of "floating voters" without binding LDP identification has made the LDP's base of voter support less stable and has forced it to be more responsive to general public policy issues.

As pointed out earlier, the LDP has used public policy successfully to deprive opposition parties of easy targets for attack. Because of the lack of strong party identification on the part of floating voters, Type 4 transactions are the least stable or predictable of the four. Even if corrective policies are taken, there is no guarantee that the LDP will receive the requisite votes to stay in power. The LDP cannot count on floating voters to consummate the political transaction. Nevertheless, the party has no choice but to hold to its end of the exchange. It must hope that the state of the nation is satisfactory and that the alternatives to LDP dominance are sufficiently unattractive that fence-sitting voters will choose to keep the LDP in power. As there is no certainty that Type 4 transactions will work out, the other three types gain in importance in the eyes of the LDP.

The four types of political exchange can be viewed hierarchically in terms of levels of politicization and the closeness of ties among the LDP, interest groups, and bureaucratic agencies. Type 1 is the most politicized, with the closest adhesion (*yuchaku*) among the LDP, interest groups, and ministries. Politicization and adhesion are a bit weaker in Type 2. Type 3 is significantly freer of politicized interference; relations between members of this interest configuration are at arm's length. Type 4 is the loosest, most diffuse, and least stable of all the configurations.

This typology of political exchange suggests that most ministries—with the notable exception of MITI, MOF, and the Economic Planning Agency (a ministry without much power)—are politicized, vulnerable to LDP interference, and penetrated in varying degrees by the very groups they are supposed to regulate. A crude ranking of ministries according to the degree of their politicization would be almost the inverse of a power-and-prestige hierarchy. The most heavily politicized ministries are Construction, and Agriculture, Fisheries, and Forestry. Slightly less but still quite politicized are Health and Welfare, Transportation, Defense, and Post and Telecommunications. A ministry that is somewhat politicized is the Ministry for Local Autonomy. Comparatively nonpoliticized ministries are MOF, MITI (excluding the Agency for Small and Medium-Sized Enterprise), and the Economic Planning Agency. This ranking, based on each bureaucracy's susceptibility to political solutions (rather than technical solutions) to policy issues, suggests that the LDP has had a harder time determining policy outcomes in areas of monetary and industrial policy (under MITI's jurisdiction) than in local issues of public works and procurements.[37]

Segmented Policy Domains

Over more than three decades of unbroken LDP rule, a complex and compartmentalized structure of interest aggregation has taken shape from the dynamic interplay among LDP electoral imperatives, changing interest-group demands, and the scope of bureaucratic jurisdiction. Policy-making in Japan tends to break down into segmented policy domains, each with its own configuration of political alignments.[38] The welter of interest-group demands are funneled through the extant structure of political-bureaucratic institutions, where they are sorted into separate and self-contained policy arenas.

The bureaucratic division of labor imposes an orderly structure on the whole process of interest aggregation. With jurisdictional boundaries clearly delineated, each ministry jealously guards its own turf against trespassing by other government agencies. Unless a policy issue happens to fall across functional boundaries, one ministry is usually given principal responsibility for formulating public policy. Not only is it unusual for other ministries to trespass; producer groups outside the policy domain (but whose interests may be affected) also have a hard time crossing over and influencing outcomes.

Cross-sector, horizontal coalitions that spontaneously coalesce on the basis of common interests do not usually emerge in Japan's segmented system. On trade-related issues, for example, it is very hard for, say, the manufacturing sectors, with an obvious stake in open trade, to pressure the makers of leather goods into lifting import restrictions.[39] Cross-sector pressures cut against the grain of the established vertical ordering. Even in cases where an industry is closely linked to another, perhaps as a supplier of intermediate goods, it is hard to form horizontal coalitions. Because the barriers against encroachment are so high, a supplier of intermediate goods may decide not to take a clear position on a given issue; or if it does, its position may clash with that of its downstream customer. At the height of the United States–Japan trade negotiations in 1980–81 over VERs on Japanese autos, for example, the rubber industry took no position on the dispute; the steel industry, another major supplier, actually came out in favor of VERs, against the position of auto manufacturers. Public interest and consumer groups, trading companies, and banks, all of which had a stake in the outcome, took no position and played no role. Hence, no unified stance or effective coalition coalesced out of the automobile trade dispute.

On the positive side, the system of segmentation lends itself to an orderly policy-making process, one based on a relatively efficient specialization of functions. It provides MITI with wide leeway to administer industrial policy within its jurisdiction. In the absence of compartmentalization, consensual decision making probably could not be practiced, at least not in its present form. There would be far too many actors vying for attention. Even if multiple demands and conflicting interests could be harmonized, the process would probably take too much time, and the transaction costs would undoubtedly be very high. Policy segmentation is thus a necessary condition for the smooth functioning of Japan's consensual system. It also appears to be a natural outgrowth of single-party dominance and interest-group inclusivity.

Mancur Olson has argued that the capacity to pursue collective and public interests is perhaps the most positive by-product of political inclusivity. In most other political systems, the dynamics of special-interest politics give rise to exploitation and ossification as individual groups gain control over wider and wider domains of policy-making. Over time, the system extracts progressively heavier costs on market efficiency. As interest groups learn how to use the system to further their own narrow self-interest—a form of learning by doing—economic collusion and political cartelization become more likely. The only countries that have successfully sidestepped this pitfall function, like Japan, under inclusive coalitions capable of overriding particularistic interests, or at least keeping them sufficiently contained within narrow policy domains that the harm done to the collective good is not allowed to metastasize.[40] "A party whose clients comprise half or more of the society naturally is concerned about the efficiency and welfare of the society as a whole," writes Olson, "particularly in comparison with lobbies for special-interest groups and congressmen accountable only to small districts."[41] Hence, the interest inclusivity of Japan's structure of single-party hegemony is one of the major institutional or regime variables that help to explain the country's capacity to carry out industrial policy.

On the negative side, however, the separation into semi-isolated policy domains has led to the protection of economically inefficient sectors, hidden transfer payments in the form of subsidies and protection to hard-core LDP support groups,[42] suboptimal allocation of capital and labor, politicization of Type 1 and 2 policy issues, foot-dragging in the resolution of Type 1 and 2 trade disputes, political favoritism, corruption, the tyranny of the status quo and status quo–based interests, and some discontinuities and contradictions in Japan's overall industrial policies (including the primary sector). Looking at it from the aggregate perspective, instead of focusing on MITI's sphere of industrial policy, we can now understand why the overall picture lacks constancy, coherence, and effectiveness, as pointed out in Chapter 1. Industrial policies for Type 1 and 2a sectors are heavily politicized and economically inefficient, posing a sharp contrast to industrial policies for Type 3.[43]

Perhaps the most serious drawback of segmentation surfaces when complex issues cannot be handled within a single policy domain. When that happens—as it frequently does in the realm of high technology—the scramble to protect bureaucratic turf can lead to policy-making paralysis. Ministries are liable to get caught up in a fight over turf and lose sight of what lies in the industry's and economy's best interests. This may be one of the biggest institutional bottlenecks in what is otherwise a comparatively bright future for high-technology industries in Japan.

In high-tech endeavors, the complexity, scope, and pace of technological change have rendered many of the old bureaucratic boundaries obsolete. The convergence of computer and communications technology, for example, has made the administrative division of labor between MITI and the Ministry of Post and Telecommunications problematic for the telecommunications, computer, semiconductor, and software industries. The two ministries, waging an ongoing battle for control of the burgeoning telecommunications industry, have found it hard to agree on a coherent policy for such pressing issues as value-added networks (VANs). Similar struggles over administrative jurisdiction are likely to arise in biotechnology, which falls uneasily into the territories of several bureaucracies—MITI, Science and Technology, Health and Welfare, Education, and Agriculture, Fisheries, and Forestry.

If rival bureaucracies cannot work out their differences, the danger of stalemate can be avoided through intervention at the top by the prime minister and his cabinet or by a blue-ribbon advisory council. Owing to the growing number of issues that fall across bureaucratic boundaries, and given the absence of viable alternatives, the prime minister and leading LDP cabinet members wind up intervening regularly. But such last-resort tactics carry a potentially high cost: namely, the politicization of complex issues that ought to be evaluated on technical grounds. Unless bureaucratic boundaries are redrawn to take account of constantly changing technologies, complex problems of high technology may be increasingly vulnerable to politicization. Standing at the edge of a new era of high technology, Japan's bureaucratic division of labor—heretofore geared to smokestack catch-up—may require some sort of restructuring.

Restructuring old patterns is never easy, given the energy required simply to overcome inertia and entrenched interests. Since the Japanese have shown themselves to be capable of institutional adaptation in the past, however, the chances that the necessary adjustments will be made are not at all remote. If old boundaries can be redrawn, there is no reason that Japan's political system cannot continue to serve as a major national strength. Certainly, the basic features of the political system can provide a stable framework for the administration of industrial policy, just as they did for the smokestack sectors during the period of latecomer catch-up.

Built-in Flexibility

The segmented political configuration model provides a parsimonious but powerful explanation for the varied patterns of policy-making in Japan. It highlights the system's strengths—the high degree of routinization, the low level of politicization in the Type 3 policy domain, and political-bureaucratic inclusivity—as well as its weaknesses—the politicized protection of noncompetitive sectors, the built-in dangers of power abuses, and the difficulties of smooth interbureaucratic coordination. To avoid the mistaken impression that the policy-making system is locked in by the bureaucratic–LDP–interest group symbiosis, it should be pointed out that there are mechanisms for significant "discontinuous change," or policy redirection, if the circumstances warrant it.

Although the LDP is dependent on its grand coalition of interest-group supporters, the coalition is so encompassing that the party can hardly be considered the puppet of any single interest group, even agriculture. Indeed, the inclusivity of the LDP's ruling coalition is such that the interests of some sectors—such as Types 1 and 2— are bound to clash with those of others—Types 3 and 4. It falls to the LDP to make the hard policy choices between the competing claims of the different sectors. From a strictly economic point of view, one could argue that the LDP has tended to let those choices be made by default—simply by caving in to particularistic, Type 1 and 2 interest pressures. Nevertheless, the LDP's propensity to follow the simple path of political expediency should not be confused with the system's capacity to rise above the pull of particularistic exchanges. Japan's hegemonic party system still retains the capacity to discipline interest groups when intolerable harm is done to the common good.

It retains this capacity in part because most of the interest groups in the LDP's inclusionary coalition have nowhere else to turn. If the LDP refuses to respond to their demands, they cannot threaten to shift their support to the opposition camp. None of the opposition parties, much less a progressive alliance, offers much hope of seizing the reins of power. Type 1 and 2 interest groups are, in a sense, hostages to the LDP. Nor is the LDP slavishly bound to adopt only those policies that please its traditional support groups.

Japan's hegemonic party system thus provides the leverage necessary for the LDP government to transcend the powerful pull of LDP– interest group exchanges and to make hard decisions that run counter to the parochial interests of even its loyal Type 1 and 2 supporters. To do this, however, usually requires the formulation of an acceptable long-term plan, sufficient time to make incremental changes, and adequate compensation for the damage sustained by clientelistic interest groups in the form of subsidies or other "side payments."

LDP Factions: Decentralized Power

If the LDP has managed to hold together an encompassing coalition, and if it has monopolized the reins of parliamentary power for so many decades, why has it not come to dominate all spheres of public policy-making? Why has MITI retained so much control over industrial policy? One would expect that where money, votes, or influence are available, the party in power would figure out a way of seizing them. One would assume that party politicians would try to maximize their own electoral interests. Moving forcefully into the policy-making arena would be one way of extending the LDP politicians' access to and control over the financial and electoral instruments of power.

This chapter and the last have already touched on a few of the reasons that the LDP has not been able or willing to bring MITI under harness—the electoral overrepresentation of agriculture, MITI's ability to reach consensus with industry, characteristics of Japanese industrial organization, and the nature of the political exchange between the LDP and Type 3 producer groups. But other facets of Japan's political regime, such as LDP factions, also enter the picture. LDP factions have helped to prevent the dictatorial rule and tight party control that might have resulted from the long history of single-party dominance.

LDP Factions

In spite of its long reign of power, the LDP is not tightly knit or organizationally strong.[44] Lacking a broad base of grass-roots support, with only 2.57 million dues-paying members, the LDP has had to rely on individual Diet representatives to mobilize voter support through the creation of their own personal organizations, called *kōenkai*, in rural and semiurban districts.[45] The LDP is often described as a federation of semiautonomous factions, constantly jockeying for power and advantage within the party.[46] In 1986, there were six factions, ranging in size from former Prime Minister Tanaka's, the largest, with 107 members, to the smallest with only 13.

Factions arise out of a combination of conditions, perhaps the most important being Japan's unusual system of multicandidacy election districts. In each district represented in the Lower House, anywhere from three to five seats are open, with each registered voter holding the right to cast only a single vote. This system places the political parties in a dilemma. Each party must decide how many candidates to run in a given district, taking into account not only its base of hard-core voter support relative to the number of seats available, but, more important, its best "guesstimate" of how the vote will break down for each of its candidates. If the LDP errs in its campaign strategy, one candidate may wind up garnering the lion's share of the votes, far in excess of the minimum number needed to win, while other LDP candidates fail to win enough to secure a seat. Or if the LDP runs too many candidates in one district, it might fracture the party's support base in such a way that no LDP candidate wins.

The system is one that places a premium on careful campaign strategy. What each party tries to do is cut down its percentage of "wasted" or "dead" votes—that is, votes cast for party candidates who are not elected. Unlike the Federal Republic of Germany's system of proportional representation, Japan's multiseat electoral system lends itself to potentially large disparities in the total votes received by a particular party compared to the total number of seats won. This disparity in the ratio of votes to seats is, in fact, one reason the LDP has managed to maintain itself in power for so long even though it receives much less than 50 percent of the votes cast.

In the rural districts, which, as pointed out already, overrepresent the farm population, only 20 percent of the votes cast are "dead." The LDP and JSP have a lock on the majority of rural and semiurban seats, making it very difficult for the JCP, Kōmeitō, and DSP to expand their bases of support.[47] The largest number of dead votes are cast in the metropolitan districts, which are underrepresented in the Lower House. These are the districts where the JCP, Kōmeitō, and DSP are strongest. But even in the metropolitan districts, the opposition camp strongholds, the LDP can count on winning its share of seats and still keep the percentage of wasted votes within a tolerable range by carefully limiting the number of candidates it chooses to run.

Japan's multicandidacy system thus gives rise to a situation in which LDP candidates wind up running as aggressively against each other as against candidates from other parties. It cannot help but foster fierce intraparty competition. Out of the crucible of this intense struggle for power emerges the phenomenon of the LDP factions. Not all parties operating within the same multiseat constituency system face factional problems; the JCP and Kōmeitō, for example, have been able to avoid LDP-type factions owing to the greater degree to which they are centralized party organizations. The multiseat, single-vote constituency system is but one structural factor, albeit a central one, related to the phenomenon of LDP factions.

Supporting the hypothesis that LDP factions are structurally related to the multiseat constituency system is the fact that there are very few districts that elect more than one member from the same LDP faction. In 1984, only ten districts out of the 130 in the Lower House, or less than 8 percent of all constituencies, had two representatives hailing from the same LDP faction.[48] LDP factions compete fiercely against one another but manage to avoid internecine warfare by staying out of precincts where fellow faction members have already established a strong base. The idea is to expand at the expense of rival factions and opposition parties, not at the expense of fellow faction members. Indeed, multiseat districts are the central battlegrounds in which the ongoing struggle between rival factions is carried out.

In order to become the prime minister, a politician usually has to be boss of a major faction, someone capable of collecting enough financial contributions to sustain his own faction and of securing enough intra-party support (in exchange for promises of cabinet and party posts) to win the party nomination.[49] This narrows the field to a small handful of candidates—less than a half-dozen in theory, but normally no more than two or three in practice—which makes the succession process less chaotic and more predictable than might be the case otherwise.

Although the existence of factions decentralizes and disperses power, it also concentrates power among a small handful of faction bosses. These leaders are capable of making on-the-spot decisions that have a crucial bearing on the country's well-being. When the party or nation faces an urgent problem, it is perfectly acceptable for a small elite of faction leaders and elder statesmen to circumvent standard operating procedures and make quick decisions; extraordinary circumstances can warrant extraordinary steps. From MITI's point of view, the concentration of power is an advantage when it expedites the often unwieldy and almost always time-consuming processes of consensus building within the LDP.

Factions are thus a mechanism for orderly competition within the accepted rules of the game.[50] Just as organizational ties structure Japan's market economy, so too human relations (*jinmyaku*), informal networks, factions, and seniority criteria impose order and predictability on political competition within the LDP. As in the case of extra-market institutions in the private sector, the imposition of organizational mechanisms (factions) on the freewheeling political marketplace has had the curious effect of intensifying competition on the one hand, while curbing its excesses on the other. It has also opened up opportunities for lateral entry from the outside while reinforcing the power of the status quo. By breaking down the undifferentiated mass of LDP Diet members into smaller subunits, factions serve to decentralize the party hierarchy, disperse power, and diversify the channels of communication.

In some respects, of course, internal decision making is rendered more difficult and time-consuming by factional decentralization. However, LDP factions should not be viewed as rigid, mutually exclusive units; if they were, the party would be hopelessly fragmented and incapable of acting in unison to achieve collective goals. Like keiretsu groupings, which are sufficiently loose-knit to permit considerable interaction across keiretsu boundaries, LDP factions leave ample slack for individual members to cross functional boundaries and forge close ties with members of other factions.

Policy Leeway

Although faction members are expected to close ranks during intraparty elections or after a party consensus has been reached, they are not committed to a particular ideology or to a clearly defined set of policies. Most LDP Diet members belong to one or more of the large number of informal policy study groups that exist within the LDP and bring together members from all factions. Because factions do not divide the conservative party along ideological or policy lines, differences of opinion on specific issues can be expressed without threatening to tear the fabric of factional unity. Functioning within a consensual system, free from ideological cleavages, LDP factions have helped to infuse the party with a healthy dose of pluralism, which has helped the party escape the extremes of iron-handed rule or stultifying stagnancy.

The LDP's policy flexibility is safeguarded by the multiseat constituency system, which tends to inhibit campaigning based on a clear delineation and differentiation of policy positions. LDP candidates tend to be afraid of advocating policies that are very controversial or explicit. To campaign on an explicit and detailed policy or ideological platform might alienate voters who do not share their views; or it might leave the candidates vulnerable to attack by competitors from rival LDP factions who speak as ambiguously as possible on controversial issues.[51] If public opinion is normally distributed, staking out a vague position somewhere in the middle of the road may be the optimal strategy.[52]

Japanese voters, especially in rural and semiurban constituencies, do not appear to demand policy specificity, much less ideological clarity, conceptual coherence, and rhetorical polish. They seem to place as much weight on the candidate's personal traits as on the appeal of policy proposals.[53] This combination of circumstances—fierce intra-party competition, factional divisions, an aversion to controversial issues, and a general preference for the middle of the road—tends to diminish the LDP's chances of molding public opinion and giving direction to public policy. Instead of setting national goals and prior-ities or at least articulating noble, universal values, the LDP is more comfortable simply reacting to particularistic interest-group demands, riding the crest of public opinion trends, and responding in ad hoc fashion to public policy needs as they arise.

Accordingly, almost by default, the LDP lets the bureaucracy as-sume the responsibility for formulating an overarching, long-term vi-sion. The bureaucracy's assumption of the role of national goal-setter runs contrary to Max Weber's notion of a natural division of labor between politicians and bureaucrats—with political leaders defining the goals and bureaucrats coming up with the technical means of achieving them. It more closely resembles the notion of a "pure hybrid" or a fusing of the bureaucratic and political roles described by Aberbach, Putnam, and Rockman.[54] Here is yet another reason that the LDP, despite its unprecedented period of Diet dominance, has not taken control of the making of Japan's industrial policy.

The fact that LDP factions are not based on ideological or policy differences but function instead as mechanisms for the organization of internal competition is significant from the standpoint of bureaucratic autonomy. This characteristic not only has weakened the LDP's ability to play the dominant policy-making role; it has enhanced MITI's ca-pacity to determine the direction of industrial policies. If MITI offi-cials can win allies among a few key faction leaders (a disproportion-ate number of whom are ex-bureaucrats), their chances of bringing about a party consensus are significantly improved. Moreover, the very existence of factions gives MITI officials plenty of room to ma-neuver.

Discontinuous Policy Shifts

Shifts in factional alignments also tend to create opportunities for major changes in policy, if they are linked to the emergence of a new prime minister. Indeed, under Japan's hegemonic party system, intra-party shake-ups involving fundamental factional realignments and the establishment of a new prime minister's administration present an ideal opportunity for the LDP government to make major, discontinuous policy changes, even those that might be politically painful. Such shake-ups are Japan's functional equivalent of the policy reorientation that takes place in Western democracies through the alternation of political parties in power. Incremental adaptations, including significant adjustments in industrial policy, can be handled by the bureaucracies on a routine, day-to-day basis. Monotonic movements, or policy redirection, usually require a shift in the factional balance of power, cabinet reshuffling, or the emergence of a new administration.

Perhaps the best example of a politically difficult policy reorientation was the implementation of a major, multiyear program called "administrative reform" (gyōsei kaikaku) in 1980. This program was intended to prune the proliferating apparatus of public administration and drastically cut public expenditures, with the ultimate aim of eliminating budget deficits that had reached staggering levels of 6–7 percent of GNP. In 1980, facing a fiscal crisis rapidly running out of control, Yasuhiro Nakasone, then the cabinet minister of the Administrative Management Agency, launched the program of administrative reform.[55] A blue-ribbon panel on administrative reform, referred to as Rinchō (Rinji Gyōsei Chōsakai, or Provisional Committee on Administrative Reform), was created in 1981, headed by Toshio Dōkō, honorary president of Keidanren. The panel consulted widely with affected groups, gave visibility and legitimacy to the undertaking, and came up with a specific set of policy recommendations. When Nakasone became prime minister following a realignment of LDP factions, he reaffirmed his commitment to the reform program by elevating it to the status of highest priority in his new administration.

The committee on administrative reform did not shy away from the hard tasks. By seeking to rationalize the country's administrative apparatus and reduce fiscal deficits, it took on some of the most inefficient but politically powerful and institutionally entrenched interests in the country: the deficit-ridden national railway system, the expensive health care system, the foodstuffs price-control system, local governments, public corporations, and even agencies of the central bureaucracy. In spite of powerful countervailing political pressures from these and other groups targeted for reform, many of the panel's recommendations were implemented. This in itself, given the political power of entrenched interest groups, was a remarkable achievement. If, as planned, the deficit is actually brought under control, the program of administrative reform will go down in history as a political landmark of sorts, achieved by power realignments within the same party that had presided over the growth of deficits to such high levels.

By eliminating deficits, however, the LDP-led government may limit the use of a major tool that it utilized to buttress its position in the past: namely, heavy public expenditures financed largely by deficit-covering government bonds. If so, this could weaken the LDP's ties with Type 2 interest groups—local construction firms and real estate interests—which depend so much on public works. If the LDP does not want to do that, it could try to reduce the budget deficits by raising the level of tax revenues, an alternative that might be resisted by other interest groups and the public at large (Type 4). Or, as is often the case, the LDP may try to diffuse or collectivize the costs by doing a little of both, cutting back public works and raising taxes. When it comes to collectivizing costs, it should be pointed out again that the negatively affected interest groups in the LDP's inclusionary coalition have no alternative party to which to shift their allegiance. Although the cutback in public expenditures may strain ties between the LDP and Type 2a interest groups in the short run, it may turn out to strengthen the LDP's position in the long term, if it restores public finances and the economy to a sound footing. Sound fiscal and monetary management is essential over the long run to limit the defection of Type 4 supporters.

The program of administrative reform is interesting because the LDP government somehow managed to implement technical solutions to seemingly intractable political problems, forcing painful sacrifices on the very political interests (Types 1 and 2) on which it depends to remain in power. There are very few instances in which a party in power has had the will or capacity to administer policies striking directly at the self-interest of groups that form the core of its coalition of support. Several factors made this unusual feat possible: (1) collective recognition of a severe problem—namely, a soaring national debt approaching potentially ruinous proportions; (2) the strength of collective norms to curb selfish, short-term interests so as to meet the national crisis; (3) a shift in factional alignments within the LDP underpinning the rise of a new prime minister and cabinet and creating the opportunity to make major changes in past policies; (4) the LDP's hegemonic position and the lack of a viable opposition party alternative for interest groups in the LDP's grand coalition; (5) the creation of an independent, neutral, highly visible blue-ribbon committee to study the problem, consult widely, mobilize a consensus, make policy recommendations, and win the public's understanding and support. The last factor, the creation of an independent advisory committee, was symbolically important. The committee provided a mechanism to transcend petty politics and bureaucratic conservatism; it also drew public attention to the urgency of the undertaking and lent legitimacy to the effort that the LDP by itself could never have achieved. By winning public support, Rinchō was able to override much of the political resistance put up by vested interest groups.

The importance of the fourth factor, the hegemonic structure of Japan's party system and the dependence of interest groups on the LDP, deserves emphasis. It is the LDP's dominance that has given structure to its interactions with interest groups and the bureaucracy. Japan's hegemonic party system has provided the overarching framework within which the government could reform the country's whole administrative apparatus; it could even reach into and revamp the deeply entrenched, politicized enclaves of economic inefficiency.

True to Japanese form, the program of administrative reform was more than the outgrowth of purely technical proposals submitted by Rinchō. It was the product of a negotiated compromise between the pressing need to change very costly, highly politicized public policies and the embedded interests of powerful interest groups like the organized workers of national railways. But the negotiated settlement does not diminish the significance either of the reforms themselves or of the system's extraordinary capacity to carry them out.

Crisis Management and Damage Limitation

The second factor listed above—norms that curb selfish, short-term behavior and encourage the achievement of long-term, collective goals—was also of great importance, since it established the primacy of the common good and lent legitimacy to the reorientation of basic policies. The strength of such norms underlies Japan's unusual ability to pull together in the face of national adversity. It disposes the Japanese to accept sacrifices, share costs and risks, and diffuse the burdens of crisis management.

Examples of collective cooperation in the face of national crisis abound: the dramatic reduction in Japan's dependence on imported oil as a percentage of its overall energy consumption following the two oil crises, antirecession and rationalization cartels, the phasing out of excess plant capacity in structurally depressed industries, and compliance with demands for voluntary export restraints. In virtually all cases, Japanese producers had to be willing to agree to a formula for the diffusion of costs, based on complicated calculations of the market share held by each corporation.

One might refer to this mode of crisis management as "damage limitation," "cost diffusion," or the "collectivization of costs and risks." It is one reason Japan has been able to adapt to abrupt or long-term, external change. As objective mediator and guardian of the public good and national interest, the state plays a key role in coordinating the strategy of damage limitation. It would be much harder for private companies to reach agreement on an equitable diffusion of costs, let alone stick to that agreement, without active government involvement.[56]

The notion of downside, cost-diffusing cooperation in zero-sum situations should be differentiated from positive-sum, benefit-sharing forms of cooperation. Japan excels at the former but is not quite as successful at the latter. Apart from cooperation in national research projects (a collective good) and in such matters as industrywide standardization of parts and specifications (low-cost coordination), there are few examples of positive cooperation in rapidly growing sectors like high technology.

Such cooperation is really not needed. Companies feel confident in their ability to compete in the marketplace and sometimes resent government attempts to structure competition. In 1972, for example, computer companies fended off MITI's efforts to organize them into specialized clusters; similar attempts by MITI to encourage mergers in the automobile and petrochemical industries also failed, owing to private-sector resistance. Interestingly, the government seems to have an easier time implementing strategies for damage limitation than it does for benefit sharing. This can be attributed to the private sector's unwillingness to accept constraints on its freedom of action in expanding markets.

From the standpoint of economic efficiency, the difficulty of organizing benefit-sharing forms of cooperation is probably a blessing. If the Japanese made a habit of dividing up an expanding, positive-sum pie, the incentives on which market competition hinges would be badly distorted. Competition provides the basis for allocative and production efficiency, entrepreneurship, and technological innovation. The Japanese are well aware of the costs and hazards of excessive market tampering in high-tech sectors, as pointed out in Chapter 1.

They recognize the dangers even in downside, zero-sum situations. Excessive reliance on cost cushioning can undermine incentives for risk-taking and delay structural adaptation. The fact that Japan has managed to adapt its industrial structure to continual shifts in comparative advantage[57] indicates that it has not resorted indiscriminately to damage-limiting measures. MITI has tried to be selective in their use, limiting such measures largely to unanticipated external contingencies, like the oil crisis, and to domestic problems that have reached crisis proportions, like public finance.

Former Bureaucrats in the LDP

Amakudari, the movement of bureaucrats into high-level positions in the private sector, was discussed in Chapter 3. More unusual is the migration of bureaucrats into elective office. More ex-bureaucrats can be found in the Japanese parliament than in the legislative branch of any other country. Of all LDP representatives elected to the Lower House since 1955, 21 percent have come from the higher civil service (see Table 4.2).[58] Ex-bureaucrats represent the second-largest of all occupational groups in the LDP, second only to local politicians (those hailing from elected office in prefectural or city assemblies), who as a group account for 26 percent of all LDP representatives. Considering the small number of higher civil servants leaving the bureaucracy each year, the 21 percent figure is statistically astounding.

Equally astonishing is the fact that virtually all bureaucrats moving into the legislative branch enter as representatives of the LDP, not as members of the opposition parties. Only 2 percent of all JSP Lower House representatives since 1955, 1 percent for the Kōmeitō since 1967, and 3 percent for the DSP since 1960 have hailed from bureaucratic backgrounds.[59] The inability of progressive parties to recruit ex-bureaucrats can be traced to the immediate aftermath of World War II, when progressive parties were at the zenith of their power and no one predicted that the LDP would dominate for such a lengthy period. The ex-bureaucrats did not simply gravitate to the LDP because it happened to be the party in power. There were other factors at work.

In 1947, when the JSP held the reins of a coalition government for only a momentary interlude of less than a year, JSP leader Suehiro Nishio tried in vain to persuade Eisaku Satō, Hayato Ikeda, and other retiring civil servants to run for the Diet as Socialist candidates. Instead, Satō, Ikeda, and the others decided to run as candidates of the Liberal and Democratic parties. Even though the JSP was in power and socialist doctrines were in vogue, the higher civil servants running for office viewed the JSP as rigid, doctrinaire, and closed—in short,

TABLE 4.2

Occupational Background of Members of the Lower House
(percent)

		LDP	JSP	Kōmeitō	DSP	JCP
All	N	732	362	98	90	90
	University graduate	0.74%	0.44%	0.49%	0.59%	0.54%
	University of Tokyo	0.29	0.10	0.02	0.13	0.20
	Local politician	0.26	0.26	0.46	0.33	0.14
	High bureaucrat	0.21	0.02	0.01	0.03	0.00
	Businessman	0.15	0.03	0.00	0.04	0.01
	Staffer	0.11	0.01	0.02	0.04	0.00
	Lawyer	0.04	0.03	0.02	0.05	0.21
	Party/organization	0.01	0.14	0.32	0.10	0.34
	Union official	0.00	0.37	0.01	0.27	0.10
1983	N	258	111	58	37	26
	University graduate	0.83%	0.44%	0.57%	0.73%	0.69%
	University of Tokyo	0.29	0.04	0.03	0.16	0.31
	Local politician	0.26	0.33	0.36	0.35	0.23
	High bureaucrat	0.20	0.00	0.02	0.03	0.00
	Businessman	0.16	0.01	0.00	0.05	0.00
	Staffer	0.19	0.02	0.02	0.05	0.00
	Lawyer	0.03	0.03	0.03	0.00	0.31
	Party/organization	0.01	0.07	0.41	0.11	0.23
	Union official	0.00	0.44	0.30	0.00	0.07

SOURCE: Based on Haruhiro Fukui, "The Liberal Democratic Party Revisited," *Journal of Japanese Studies* 10, no. 2 (1984): 393.

NOTE:

All: All members of the respective parties who have ever been elected to or served in the Lower House since the founding of their party: 1955 for LDP, 1955 for JSP, 1967 for Kōmeitō, 1960 for DSP, and 1946 for JCP.

1983: Members elected in the Dec. 1983 general election.

Local politician: Former elected local politicians: overwhelmingly former prefectural assembly-men.

High bureaucrat: Former senior officials in central government ministries and agencies.

Staffer: Formerly on other politicians' personal staffs.

Party/organization: Former or current functionaries of a political party or organization.

not a party suited to their tastes or hospitable to their political aspi-rations. The JSP's reliance on labor unions and the importance placed on union experience for rising within JSP ranks also diminished the civil servants' prospects for upward advancement. Not only did they sense a basic incompatibility; they also did not foresee a particularly promising future for the JSP and other opposition parties. If they were going to enter the world of politics, they reasoned, better the Liberal or the Democratic party (prior to the merger of the two into the LDP, as it is known today) than other parties.[60]

For the LDP, the large influx of ex-bureaucrats began in the late

218

The Politics of Industrial Policy

1940's, when Prime Minister Shigeru Yoshida recruited many of the country's top retiring civil servants including Hayato Ikeda, Eisaku Satō, Takeo Fukuda, and Masayoshi Ōhira—all of whom later became prime ministers. In the 1949 election, more than 50 former bureaucrats were persuaded to run for Lower House seats; more than 30 of them were elected.[61] Many joined the Yoshida faction, forming the nucleus of what became known as the "Yoshida School" (Yoshida gakkō). Consisting of the colorful leader's many disciples, the Yoshida School became the LDP's "conservative mainstream" (hoshu honryū), providing Japan with the leadership it needed during the most formative period of its postwar reconstruction. The Yoshida School can take much of the political credit for laying down the basic postwar policy direction—rapid economic growth, limited rearmament, and close relations with the United States—that has brought Japan unprecedented peace and prosperity.[62]

For the JSP, the most common occupational path to elective politics is through organized labor. A large proportion (37 percent) of all JSP Lower House representatives since 1955 have come up through the labor union ranks.[63] The contrasting patterns of LDP and JSP recruitment—one from the higher civil service, the other from organized labor—could hardly be more stark. The DSP similarly draws on candidates with labor union backgrounds. More than a quarter of its representatives have served in private-sector labor unions (which tend to be less militant than the public-sector unions that support the JSP). For the JCP, only 10 percent have come up through the ranks of labor union organizations, with none from the higher civil service, and 34 percent from within the Communist party itself. Similarly, 32 percent of Kōmeitō representatives have risen from within the party itself or from the Sōka Gakkai, the religious organization with which the party is closely affiliated. Not surprisingly, the JCP and Kōmeitō, the two parties with the highest percentage of internal recruitment, happen to be the most centralized, tightly organized, and probably least open to outside entry. The one route of recruitment to the Lower House common to all parties is local government, which is either the largest or second-largest occupational category for all parties except the JCP (see Table 4.2).

The contrast between LDP ex-bureaucrats and JSP ex–labor union officials is even sharper in the Upper House, where national constituencies account for 49 out of 125 seats. To win in the national constituency, as opposed to local districts, Upper House candidates often need the backing of organizations that have national networks. Several ministries and labor unions can offer candidates such networks. Typically, over 40 percent of all LDP members in the Upper House come from bureaucratic backgrounds, whereas over 60 percent of all JSP members come from labor union backgrounds.[64] This suggests that, in terms of personnel, the two largest parties pose so sharp a contrast—the higher civil service versus the public-sector labor unions—that it comes as close as any factor to constituting a class distinction. The distinction reflects underlying cleavages in the network of groups with which the two leading parties are closely aligned.

The impact of the LDP's successful recruitment of ex-bureaucrats has been far-reaching. No doubt it has extended the LDP's longevity as the majority party in power. The higher civil service has served as the LDP's most fertile source of leadership talent. Most of Japan's postwar prime ministers have come from the higher civil service, and a disproportionate percentage of the most influential faction leaders have also come from bureaucratic backgrounds.[65] Moreover, looking at all cabinets from 1948 to 1977, we should note that 183 out of 425 cabinet members, or an amazing 43 percent, have been ex-bureaucrats. That nearly half of all LDP cabinet members have been ex-bureaucrats is an indicator of the disproportionate power they have wielded within the ruling conservative party.[66]

Because entrance to the higher civil service is based on meritocratic criteria—specifically, a very rigorous set of examinations—the bureaucrats who have won seats in the Diet represent the country's "best and brightest." Quite apart from native ability, they have infused the conservative party with a high level of leadership ability, firsthand experience in public administration, intimate knowledge of the policy-making processes, access to the best available information, and extensive contacts with leaders in both the public and private sectors.

It is hard to imagine how the LDP might have fared without the continual infusion of new bureaucratic talent. What would have happened if the opposition parties had managed to attract their fair share? Although this is, of course, pure speculation, the likelihood is that Japan's postwar political history would have turned out very differently.[67] LDP leadership would not have been as effective, nor would the conservative party have been able to stay in power for so long. The LDP's unchallenged capacity to draw top talent from the higher civil service has made a significant difference.

Specifically, the large influx of higher civil servants has led to the consolidation of ties between the LDP and the various ministries. This marriage has turned out to be one of the LDP's greatest assets. It has joined the bureaucracy's administrative skills and technical know-how with the LDP's constitutional authority and political mandate, a union that has led to comparatively effective public administration and a very stable structure of governance.[68] With so many former bureaucrats in the fold, the LDP is able to maintain extensive channels of communication with all ministries, especially those that are functionally crucial (like Finance) or politically central (like Agriculture, Fisheries, and Forestry). The LDP hears vital news before other parties and is on the receiving end of far more information. Much of it is transmitted informally through the network of personal relationships linking ex-bureaucrats to their former ministries. Being shut out of this insiders' loop makes it harder for the opposition parties to participate as fully as the LDP in the policy-making process. Moreover, without a nucleus of competent former public servants, the opposition parties have had trouble convincing the public that they have the ability to govern. Here, in image and reality, is an example of the extraordinary advantages that have accrued to the LDP as a consequence of its recruitment patterns.

It is often asserted that the recruitment of higher civil servants strengthens the hands of the administrative branch more than it does the ruling conservative party because it places loyal ex-officials in strategic posts within the LDP. The notion is that the "old boys" are sympathetic to the views of the ministries in which they once served and try to build support for those views within the LDP. This belief contains more than a grain of truth. However, one must be careful to avoid the mistake of posing zero-sum questions and reaching overly simple conclusions. The question of which side benefits more—the ministries or the conservative party—is, in some senses, a misleading one to ask. The overriding fact is that both sides benefit immensely. It is not an either-or situation.

Of the ministries represented in the LDP, the Ministry of Finance (MOF) has the most members with 30. The reason MOF sends so many into elective office is not simply a by-product of its standing as one of Japan's most powerful ministries. MOF happens to have a natural political springboard built into its organizational structure: namely, the national network of tax offices located throughout Japan's regional prefectures.[69] This network places MOF officials in a position to develop key contacts with local leaders, business firms, and special-interest groups that can be converted subsequently into electoral support. MOF also possesses the organization to mobilize votes for national constituencies in the Upper House, as pointed out earlier.

In view of MOF's organizational reach, neither the high proportion of MOF officials in the LDP nor their prominence in electoral politics is at all surprising. Several of Japan's best-known and most able prime ministers—Ikeda, Fukuda, and Ōhira—have come from MOF.

Other ministries that are well represented within the LDP include Type 1 and 2 agencies, such as Agriculture, Fisheries, and Forestry, Construction, Transportation, Health and Welfare, Local Autonomy, and Post and Telecommunications—the ministries involved in the closest and most politicized relationship with the LDP. Here again, each of these ministries offers a local or national springboard (or both) for officials entertaining ambitions to run for elective office. The presence of alumni from Type 1 and 2 ministries has helped the LDP consolidate ties with special-interest support groups and extend its control over pork-barrel exchanges.

Until quite recently, MITI had not sent many of its officials to the Diet. It could count on a few highly placed alumni in the LDP, like Etsuzaburō Shiina, former deputy prime minister, to watch out for its views and interests. But it had neither MOF's built-in springboards nor the political exchange relationship with the LDP that Construction, Agriculture, Fisheries, and Forestry, Transportation, and Post and Telecommunications shared. MITI's officials had an array of posts outside politics from which to choose lucrative second careers. For many years, MITI felt little need to encourage its officials to seek elected office. Indeed, the idea of working in the world of politics (*seikai*) carried distinctly unsavory overtones for some MITI officials who disdained political interference in the formulation of industrial policy.[70]

The inclination to keep some distance has changed, however, owing to the decline of MITI's power and the LDP's encroachment on what had been the bureaucracy's terrain. Since the mid-1970's, an unofficial and largely unspoken conclusion appears to have been reached at MITI: that placing more officials in the Diet would be advantageous. Increasingly, MITI officials have begun to leave MITI at a younger age to run for office. Even without a natural springboard or national organization, MITI has succeeded in placing more of its alumni in the legislature. As of 1988, seven MITI "old boys" held seats in the parliament: six in the Lower House and one in the Upper House.

If, as anticipated, the trend toward a slowly expanding role for politicians continues, MITI cannot help but benefit from having a cadre of its own alumni strategically placed within the LDP. This will help to assure support for budgetary allocations and legislation directly related to MITI's industrial policy. It may also facilitate MITI's desire to extend its influence over policy issues that currently fall across bureaucratic boundaries. Insiders say, for example, that MITI, quietly and behind the scenes, was able to have some influence over the reform of Japan's expensive health care program owing to the strategic placement of one of its alumni in the post of Minister of Health and Welfare.[71]

MITI's interest in strengthening its political clout is not confined to the national Diet. Since the mid-1970's, it has also begun to expand ties with a number of the country's 47 prefectural governments. MITI now sends young officials to serve in prefectural governments, often as assistants to the governor, in order to assist in regional planning for the creation, promotion, and location of industries. The extension of MITI's reach into the offices of prefectural governments is in keeping with its emphasis on regional development and on the creation of "technopolises"—industrial complexes consisting of high-technology industries—throughout Japan. It is noteworthy that a MITI alumnus, Morihiko Hiramatsu, has been elected governor of Oita Prefecture on the island of Kyushu, a region actively trying to entice high-technology companies, both domestic and foreign, to set up manufacturing and assembly facilities.

All of MITI's moves, especially the migration of more MITI alumni into the LDP and the establishment of closer links with regional governments, are clearly designed to expand the scope of its political influence. MITI has consciously chosen to make these changes in response to the decline of its own base of power and to the myriad of subtle but significant changes taking place all around it—in Japan's political system, in the nature of the international environment, and in Japan's transition from smokestacks to high technology. Operating in an environment of such fluidity, MITI, like other Japanese bureaucracies, wants to make certain that its influence does not wane. One of the surest ways of ensuring that it does not is to solidify the nexus of ties with regional governments and the LDP.

In addition to having its own alumni within the LDP, MITI is also able to call on political support from a large number of LDP representatives who are organized into various groups for the promotion of high-technology industries.[72] The groups vary in size, purpose, membership, level of formality, and effectiveness. They range from small, official, industry-specific committees like the LDP Subcommittee on Telecommunications to very informal, unorganized, ad hoc support groups (zoku).[73] The size and strength of zoku serve as indicators not only of the political clout wielded by domestic producers but also of the degree of LDP-industrial adhesion and the extent to which the bureaucracies in charge are politicized. Not surprisingly, Type 1 and 2 industries—Construction, Agriculture, Fisheries, and Forestry, Transportation, Post and Telecommunications, and so on—rank at the top.

Between the official LDP subcommittees and the highly informal zoku are several LDP caucuses (*giin renmei*), representing the largest collection of LDP lobbyists for high technology.[74] Membership is open to all LDP representatives who want to climb aboard the high-tech bandwagon. In return for support during Diet deliberations on such matters as high technology–related legislation, finance, taxes, and budgets, members of the high-tech caucuses receive financial contributions from private corporations and industrial associations, both directly and through trusted intermediaries. Since the membership fee is small, and since support for high-tech industries can be justified on the grounds of advancing the national interest, the membership in these caucuses tends to be large—over 150 in some cases—but the working core of activists is usually much smaller, somewhere between 10 and 25.

The two most prominent high-technology caucuses are the Information Industries Promotion Caucus (IIPC; Jōhō Sangyō Shinkō Giin Renmei) and the Knowledge Industries Promotion Caucus (KIPC; Chishiki Sangyō Shinkō Giin Renmei). IIPC, organized around Tanaka faction members, deals almost exclusively with the computer industry. KIPC, organized around Fukuda faction members, is in charge of all other high-technology industries: robotics, machine tools, aircraft, space, and biotechnology. The aim of the two caucuses goes beyond support for specific pieces of high-technology legislation; the goal is the creation of a business environment highly favorable to the growth of high-tech industries.

During the season for budget compilation and legislative action, the deputy directors of MITI's various divisions in charge of high technology—such as Electronics Policy, Industrial Electronics, Data Processing Promotion, and Aircraft—keep in close touch with leaders of IIPC and KIPC in order to explain the aims of MITI policies to make sure that MITI proposals will be backed by the Diet. The pattern of relying on legislative-branch support from the LDP caucuses is true not just for high technology but also for most of the big line items on MITI's budget, including small and medium-sized enterprise and energy. MITI has helped create active LDP caucuses for both of these areas, just as it has for high technology.

Exactly how much of a difference the two information industry caucuses make is not clear. The view of most MITI officials is that the support of LDP caucuses is helpful only at the margins. They cannot squeeze more out of the budget than MOF is willing to concede; and the processes of MITI-MOF negotiations over the budget hinge less on the application of political pressure than on the logical tightness of the budgetary requests submitted. The support of the two LDP caucuses for proposed legislation is also not critical; seldom does it spell the difference between Diet passage and veto. Even without the involvement of IIPC and KIPC, MITI officials believe, they would probably still get what MITI wants, particularly because high technology has become the rage in Japan.

Why, then, is there a need for the two caucuses? Why, indeed, has MITI bothered to help create them? To answer requires going back to the concept of collective insurance discussed in Chapter 1. Simply put, MITI officials want to reduce the element of uncertainty. The LDP caucuses for high technology represent a low-cost, low-risk political insurance premium. Even if the two LDP caucuses are not absolutely essential in the short run, MITI officials believe they can still turn out to be valuable assets in the long term. Who knows when they might be needed? Educating LDP representatives on the importance of high technology takes time. Starting the process early and mobilizing as many legislators as possible to the banner of high technology thus makes a great deal of sense.

At the same time, however, the existence of the two LDP caucuses raises the disturbing possibility that Japanese industrial policy for high technology might be increasingly subjected to political interference. Would it not be tempting for private corporations or industrial associations, for example, to turn to the two caucuses for help, should MITI refuse to adopt policies that they deem indispensable to their interests? If IIPC and KIPC had existed at the time of the United States–Japan negotiations about color television, for example, would the disgruntled producers of color televisions and their industrial association have turned to the two LDP caucuses for help in stopping MITI from making unilateral concessions? Has MITI created a mechanism capable of turning the Type 3 configuration of generalized political support into a Type 2 relationship of reciprocal patronage?

The danger of that happening is remote. The high-technology industries have too much at stake to maneuver behind MITI's back. Nor is MITI apt to alienate the high-technology sectors so badly that it loses their trust. Never has MITI's relationship with industry degenerated to the point where the basis for a modus vivendi has completely disintegrated. Nor would the two LDP caucuses want to act in ways that jeopardize the relationship of reciprocal trust between MITI and the private sector. So long as MITI and the private sector can continue to reach consensus, the incentives to preserve the MITI-industry marriage will remain strong for all parties. IIPC and KIPC will continue to operate, as they do today, on behalf of MITI and the high-technology industries, within the constraints imposed by the Type 3 political configurations.

The Japanese Polity

Faced with the challenge of adapting to a constantly changing political environment, in short, MITI has shown an amazing ability to maneuver adroitly. It has come up with an ingenious set of responses to the erosion of its political power, such as quietly encouraging more of its alumni to run for elective office, extending its network of ties with prefectural governments scattered throughout the hinterland, and organizing LDP caucuses for high technology. MITI's agility both reflects and reinforces the capacity of Type 3 actors to protect their policy domains from dysfunctional political intervention, one of the most important characteristics of Type 3 industrial policy for high technology.

Bureaucracies in other countries, such as the ministries of industry in Sweden and France, also provide a measure of protection against partisan political tampering; but the carapace is neither as big nor as hard as that provided by MITI. Few, if any, other bureaucracies in the world enjoy the close relationship that MITI does with Japanese industry. It is this combination of MITI-industry synergy and a tacit MITI-LDP division of labor that sets Japanese Type 3 industrial policy apart from that of other countries. What makes the combination possible is the structural support provided by Japan's political system, particularly the dominance of the LDP, the inclusivity of its interest coalition, and the nature of the Type 3 interest aggregation.

Cast within the context of Japan's political system, the whole debate over which branch of government dominates policy-making—the LDP or the bureaucracies—comes across as simplistic.[75] The processes of industrial policy-making are far too complex to be reduced to the one-dimensional question of which branch of government is in charge. Industrial policy emerges from the subtle interplay of economic, political, and bureaucratic forces in the context of one-party dominance. It is the political system that provides the institutional framework within which industrial policy is formulated and administered. The political system sets the stage, sorts out the actors, and defines the rules by which the whole process is carried out. Politics is primary.

To assert the primacy of politics (broadly defined) is certainly not to deny the power of the Japanese bureaucracy. In Type 3 policy domains, the bureaucracies do indeed exercise the greatest influence. MITI is unquestionably the dominant single actor with respect to Japanese industrial policy for high technology. Yet MITI's power is dependent on the political configurations that take shape within each policy domain. The specific configurations in high technology set the broad parameters within which MITI and the other bureaucracies are able to exercise power.

Japan was described in the last chapter as a "network state"—that is, one able to exercise power only in terms of its network of ties with the private sector. The existence of various organizational structures in the Japanese economy gives MITI numerous levers, or points of access, by means of which to intervene in the marketplace. In the same vein, bureaucratic power is also relational in the sense that it emerges from the structure of LDP-bureaucracy-interest group alignments and the political exchanges that take place among them. The secret to the power of the Japanese state is thus embedded in the structure of its relationship to the rest of society. Japan is without question a societal state.

Nearly all analysts of Japanese society, from Masao Maruyama to Robert Bellah and Chie Nakane, have stressed that the state is the symbolic and functional linchpin of society.[76] Because Japan is strongly oriented toward the achievement of collective goals, the state functions as the mechanism for mobilizing resources and coordinating groups in the pursuit of these goals. In this very broad sense, the polity takes priority over other sectors of society, including even the economy.[77] The economy is of importance in providing the instrumental means of achieving larger, collective goals set forth by the polity. The Japanese state derives its legitimacy from its capacity to coordinate industry-specific efforts and national goals.

The polity is not merely one of the main pillars of society, nor even the cornerstone. It is the essential edifice. The state is to Japanese society what unbroken generations of the household are to the nuclear family: the source of collective identity, guarantor of continuity, source of cohesion and unity, mechanism for conflict resolution, guardian of common interests, and main vehicle for reaching collective goals. The state is not merely an administrative appendage, as it is sometimes seen in the West, responsible only for protecting individual rights, establishing equitable rules and norms, and allocating resources. In Japan, the state symbolizes and functionally affirms the solidarity of the national collectivity.[78]

Perhaps the closest equivalent that can be found in the West to the concept of the Japanese state is the philosophical tradition of the "organic state," which is based on a composite of ideas stretching from Aristotle and Roman law through Thomas Aquinas and medieval natural law all the way to contemporary Catholic social philosophy.[79] The organic state is similar to the Japanese state in the fundamental assumption that the Aristotelian *polis,* or political community, serves as the basis for the polity. The *polis* takes priority over the individual as well as subordinate social units because the whole is of higher importance than any of its parts. The state's central duty—indeed, its moral obligation—is to safeguard and enhance the common good.

This obligation provides the normative underpinnings for a potentially interventionist—if not authoritarian—state. The organic state is given sweeping authority to take whatever actions are thought to be necessary to protect the common good. It has the power not only to override political opposition but also, in extreme cases, to suspend the institutional framework and the procedural and legal order within which the system operates (as has been seen in Latin America). The concept of the organic state does not necessarily rule out the possibility of governing within a decentralized market economy, functioning on the principle of maximizing private interests. As long as the market economy serves the common good, there is no fundamental incompatibility.

But the organic state's attitude toward the interest-maximizing behavior of private corporations is a far cry from that of the liberal pluralist state, founded on Adam Smith's faith in the wonders of the "invisible hand." It does not share the liberal pluralist assumption that the common good is best advanced by leaving outcomes to freewheeling economic and political competition. Because private interests are narrow, self-centered, and capable of damaging the collective interests of the political community, the organic state feels that it has not only the right but also a positive obligation to intervene in ways that rein in the behavior of self-centered private actors. Given the strength of the organic state tradition in Latin America and the Iberian countries, it is not surprising that corporatist states and authoritarian regimes have appeared there, particularly in light of their socioeconomic patterns of development and the nature of their integration into the international political economy.[80]

The concept of the organic state thus bears some similarities to the Japanese state: the primacy of the political community rather than its constituent parts, overriding emphasis on the common good, pursuit of collective goals, difficulties in establishing a basis for legitimate political opposition, and normative justification for state intervention. However, the two concepts of the state are very different in several crucial respects. The Japanese state operates on the basis of a more particularistic value system.[81] It derives much more of its power and effectiveness from its capacity to work in basic harmony with, and constructively channel the enormous energy of, other segments of society, ranging from private corporations and industrial associations to political parties and LDP support groups. The fact that the power of the Japanese state is so deeply lodged in the structure of society suggests that there are intrinsic limits on its use. Unlike the organic state, the Japanese state has no choice but to rely on consensus, habits of compliance, and voluntary cooperation on the part of private actors to get things done. It simply cannot base its rule on coercion, threats of legal sanctions, or unilateral imposition of its will on society. When the Japanese state is forced to fall back on naked coercion, as it did during the 1930's and the Second World War, there arise serious problems of political instability and the looming specter of deep-seated disorder.[82] It is hard to imagine the Japanese state ruling by coercion under normal circumstances.

On the other hand, the state's capacity to sustain a consistently favorable business environment, one that facilitates the work of the private sector, is one of the major reasons for Japan's dynamism. It is from the synergistic interplay of the state and private enterprise, working together to advance common interests and objectives, that Japan draws its energy, resilience, and adaptability. Being responsive to the ever-changing needs of the private sector—indeed, just keeping in constant touch with it—forces MITI and other ministries to continue the process of adaptation. MITI's resilience is evident in the adjustments made in Japanese industrial policy as it has shifted from heavy, old-line industries to high technology. And what has made the transition from latecomer to pioneer smoother than expected so far are the regime characteristics of Japan's political economy, including attitudes toward government intervention, characteristics of industrial organization, and above all the political framework of industrial policy-making.

Conclusions

In assessing the experiences of advanced industrial countries, a number of analysts have pointed out that the impact of industrial policy on economic performance has been mixed at best.[1] Whatever the short-term gains in terms of promoting targeted sectors or responding to the demands of politically active interest groups, the use of industrial policy has exacted a high toll in terms of breeding long-term economic inefficiency. Even the experience of postwar Japan—frequently cited as a shining example of effective central planning and state intervention—cannot be considered a history of unmixed success. On close examination, Japan's postwar record can be seen to contain some of the same elements of unevenness, inconsistency, and political expediency that have bedeviled the industrial policies of countries like Italy and England. The telltale signs can be found here and there in scattered pockets of economic inefficiency: agriculture, lumber, coal, retail distribution, food processing, and tobacco, to name just a few. In an integrated domestic and world economy, such enclaves of inefficiency have given rise to economic distortions, including the suboptimal allocation of resources and chronic trade conflicts.

As this book has pointed out, the most egregious pockets of inefficiency tend to be located in sectors that lie outside MITI's jurisdiction—especially the highly politicized, domestically oriented ministries like Agriculture, Fisheries, and Forestry. For industries subsumed under MITI's jurisdiction—including most of the key manufacturing sectors—industrial policy has been more coherent, forward-looking, and constructive. Industrial policies for steel, lasers, and semiconductors pose a sharp contrast to those for agriculture, food processing, and construction. The differences can be attributed largely to variations in the patterns of interest aggregation involving the LDP, producer groups, and bureaucratic agencies.

Looking at Japan in comparative perspective, we have noted that the policy instruments used to promote the development of high technology—namely, technology push and demand pull—are no different in kind from those utilized by other countries. To the extent that there are perceptible differences, they lie in degree and overall mix. MITI has placed greater emphasis on technology push through such mechanisms as national research projects and less on direct demand pull through public procurements. For reasons discussed in this book, the Japanese have had more success in administering key instruments of technology push, like national research projects, than the United States or most countries in Western Europe (though the advances should not be exaggerated). Hence, though the general instruments of industrial policy are the same, the blend is different and the results are often more fruitful.

No single reason can be cited to explain the greater effectiveness of MITI's industrial policy. It is not simply a result of MITI's unique strengths—its superior organization and talent, the history and scope of administrative power, its network of amakudari alumni in high-level positions, and its close working relationship with the private sector. Nor can its effectiveness by attributed exclusively to the organization of Japanese capital markets—specifically, the capacity of MITI and the Ministry of Finance to allocate credit to targeted sectors. Rather, the secret to Japan's apparent success lies in the overall system within which industrial policy functions. It is the dynamic combination of factors, interacting within the structure of an integrated system, that gives shape to the apparent effectiveness of Japanese industrial policy.

The word *apparent* ought to be underscored, since neither this book nor any other of which I am aware has established (in terms of strict scientific criteria) a clear causal connection between industrial policy and competitiveness in high technology. The weight of the evidence strongly suggests that industrial policy has facilitated the development of Japan's high-technology industries, but causality has yet to be proven. From the standpoint of scientific method, the connection may be only correlational or perhaps even spurious in nature. It is hard to demonstrate beyond a reasonable doubt that the instruments of industrial policy have given Japanese companies their competitive edge. Japan's early advances in high technology may have had less to do with the content of industrial policy than with the intrinsic strengths of the private sector, or perhaps the soundness of macroeconomic policies.

If industrial policy cannot be given conclusive credit, then what can be said about its role? Quite apart from its impact on specific sectors, one can argue that industrial policy has functioned as an indispensable mechanism in the machinery of Japan's political economy. It has served as the main instrument for consensus building, the vehicle for information exchange and public-private communication. Close government-business relations would be hard to imagine in its absence. Indeed, the whole system of consensus, on which Japan's political economy relies, would be hard to maintain without industrial policy as an integrative mechanism.

From the standpoint of bureaucratic politics, we should also note that MITI uses industrial policy as a lever to strengthen and extend its power vis-à-vis rival ministries. MITI utilizes industrial policy to seize the initiative on important issues, like the development of high technology. Its aggressive, promotional approach to policy-making sets MITI apart from such ministries as Health and Welfare and Post and Telecommunications, which follow a more conservative philosophy of regulatory control. This has enabled MITI to expand its influence on policy issues related, for instance, to biotechnology and telecommunications—issues that fall across jurisdictional boundaries. The ministries of Health and Welfare and Post and Telecommunications have responded recently by adopting a more forward-looking, promotional posture.

Cast against the broad background of Japan's political economy, therefore, the role of industrial policy goes well beyond addressing sector-specific needs. Even if it did little or nothing to facilitate the growth of targeted sectors (which is not the case), industrial policy would still perform the indispensable functions of building consensus, sustaining communication networks, maximizing the use of resources, and consolidating ties between government and private enterprise. Industrial policy functions as a linchpin, holding together Japan's system of consensual policy-making.

Industrial policy also serves as a lever in the continual infighting that takes place between bureaucracies in Tokyo. There is no doubt that it strengthens MITI's hand in dealing with rival bureaucracies— not to mention vis-à-vis the LDP and the private sector. In this sense, the role industrial policy plays is to counteract the structural tendency toward power dispersion in Japan, a strong centrifugal force that has plagued the Japanese state throughout its two-thousand-year history. Japan's problem with respect to political power is not the danger of excessive concentration that has bedeviled Western societies; it is, rather, the problem of finding ways to overcome the dangers of political stalemate and policy-making paralysis that arise from power dispersion.[2]

The role of general system support suggests that industrial policy is not something that can be easily abandoned, even though the era of latecomer catch-up is past and Japan's economy has matured in certain areas, like the development of capital markets. If industrial policy once functioned to compensate for market imperfections, the need to rely on it as a compensating mechanism is now significantly reduced; and if the government once derived its capacity to implement industrial policy from the allocation of substantial financial resources, that capacity has been significantly diminished. Foreign pressures to drop what are perceived abroad as unfair industrial policy practices, like antirecession and rationalization cartels, have also been stepped up. Yet even in the face of such developments, MITI is unlikely to relinquish industrial policy because it plays a central a role in the day-to-day functioning of Japan's political system. What it provides is a framework for communication and consensus building between government and business; this, in turn, serves as a buffer against excessive meddling by the political parties. In very few, if any, of the other major economies is the level of politicization associated with industrial policy kept as low.

Distinctive Factors

Viewed in comparative perspective, MITI's industrial policy has emerged from a distinctive set of regime characteristics. Of these, perhaps the most unusual are the long period of Liberal Democratic Party (LDP) domination, the weakness of labor-based political parties, and the lightness of Japan's defense burden. In few other large industrial states can one find any one of these characteristics—not to mention the combination of all three.

Of the three, undoubtedly the most decisive is the LDP's unprecedentedly long reign as majority party in power. The LDP's postwar monopoly has made the so-called LDP-bureaucratic alliance possible, featuring an exceptionally close, nonantagonistic working relationship between the legislative and executive branches of government. In the division of labor between legislative and bureaucratic branches, the LDP has tended to concentrate on servicing the interests of traditional support groups (referred to as Types 1 and 2 in this book) through fiscal expenditures, regulatory control, and pork-barrel activities of all sorts, while leaving MITI largely in control of formulating industrial policy for most of the key manufacturing sectors (in close consultation with private industry). The modus vivendi has given MITI unusual leeway to formulate a comparatively nonpoliticized set of policy measures for the development of high technology.

The weakness of labor-based parties, the reverse face of LDP dominance, has also contributed to the nonpoliticized nature and adaptability of industrial policy. In the absence of powerful labor-based parties, Japan has not had to develop a neocorporatist system of tripartite negotiations (involving labor, business, and government) concerning major economic issues, as many countries in Western Europe have. The muted power of labor-based parties has insulated MITI from the pressures actuely felt by governments in Great Britain and Italy. Unlike the British and Italian governments, MITI has not had to alter the substance of industrial policy radically in order to accommodate the demands of organized labor. It has not been compelled, for example, to accommodate labor demands for massive subsidies, import protection for declining sectors, nationalization of structurally depressed industries, ad hoc efforts to create and maintain employment in economically depressed regions, generous transfer payments for the unemployed, large-scale job retraining programs, restrictions on overseas investment, and the administration of an income policy favorable to the wage expectations of organized labor.

The comparative weakness of organized labor has had a profound effect on the postwar development of the Japanese economy. Instead of following a labor-led road to the creation of a welfare state, as most countries in Western Europe have, Japan pursued instead the fastest pro-producer route to economic growth. Only after doubling output every seven or eight years did the LDP government turn belatedly to the task of upgrading the country's welfare programs. By then, the tax revenues at its disposal had multiplied as a result of years of pace-setting growth, permitting the Japanese government to expand welfare outlays without crowding out capital investments. It could combine monotonic increases in welfare outlays with reduced but still rapid rates of aggregate growth.

Without militant demands from organized labor, Japan has also managed to escape the trap of protecting industries no longer capable of competing against low-cost producers in latecomer countries. What the Japanese government has followed instead is the politically difficult path of relying on positive adjustment policies, like the scrapping of excess plant capacity. Here again, it would be hard to name another

234 *Conclusions*

country that has cut back plant capacity instead of administering the easier short-term remedy of propping up declining industries. Nearly all countries respond to the problem of structural adjustment by keeping ailing sectors on costly life-support systems. Freed from labor-based pressures on issues involving trade, subsidies, employment, wages, welfare, and foreign direct investment, MITI has had the flexibility to formulate a comparatively nonpoliticized set of industrial policies that have minimized economic distortions.

Japan's very low level of defense spending—around 1 percent of GNP—is another unusual feature that has enhanced MITI's flexibility in designing industrial policy.[3] Japan has not had to divert highly skilled manpower resources from commercial to military endeavors. Unlike their counterparts in the United States, Japanese scientists, engineers, and research personnel have been able to concentrate almost exclusively on commercializing breakthroughs in knowledge.[4] The low dissipation rate has helped to make Japan a formidable competitor in world markets. Japanese companies have learned how to recoup upfront R&D investments through the sale of commercial products. MITI has squeezed maximal mileage out of the resources channeled into national technology-push projects; and the whole R&D system has worked with enviable efficiency in terms especially of innovations in process technology. Because improvements in process technology hold the key to cost reductions and improvements in product quality, Japan's strengths in commercial applications and process innovation have given it a decisive competitive edge. By contrast, the U.S. R&D system suffers not only from the heavy drain on manpower and financial resources generated by military demand but also from the low prestige and rewards accorded to the unglamorous but commercially critical aspects of process technology. Small wonder that Japan has caught up with and in certain areas overtaken the United States in high technology.

The combination of enormous size and remarkable social homogeneity also sets Japan's political economy apart from other nation-states. In terms of domestic demand, Japan is already the second-largest economy in the world and though its economy is only about half the size of that of the United States, Japan is slowly but steadily catching up. The sheer size of Japan's domestic market means that MITI enjoys the luxury (denied most European states) of being able to aim at the development of across-the-board strength in virtually all areas of high technology. Owing to the small size of their domestic markets, none of the European states—not even West Germany—can hope to compete in all areas of high technology. In consequence, the European states have had to pursue a niche strategy, concentrating limited resources in specialized areas where they have a comparative advantage, like highly sophisticated machine equipment and industrial chemicals in West Germany. Taken as a whole, Western Europe is large enough, of course, to sustain a presence across the high-technology spectrum; but orchestrating a communitywide strategy of comprehensive coordination has proven to be politically difficult.

The homogeneity of Japan's large population is also without parallel. Although a few countries in Western Europe, like the Scandinavian states, are as homogeneous, they are only a fraction of Japan's size. Sweden, the largest, has a population of only 8.3 million; Norway has only 4.1 million; and Denmark, 5.1 million. With over 120 million people, Japan's population is six times as large as the three Scandinavian states combined. The two Koreas come somewhat closer to Japan's size, but Japan is still more than twice as large as the two Koreas.

The absence of deep social cleavages—ethnic, racial, religious, regional, or historical—obviates the need to rely on industrial policy to solve a variety of social problems that might otherwise arise. In contrast to the U.S. government, MITI does not have to shoulder the burden, for example, of formulating, administering, and monitoring affirmative action programs. To get an idea of the potential costs that would be incurred if Japan had to deal with greater social problems, one need only refer to the economic inefficiencies caused by having to protect the meat and leather-goods industries, in which large numbers of the so-called *burakumin* (outcast) group are represented. Imagine the economic costs if the number of such societal problems were multiplied.

Structural and Sociocultural Factors

The distinctive factors discussed above do not explain fully MITI's capacity to formulate and implement a comparatively effective industrial policy; they constitute necessary but not sufficient conditions. To understand how MITI can administer industrial policy without either causing the market mechanism to malfunction or expanding permanently the state's role in the economy requires that structural and sociocultural factors be included in the analysis: specifically, Japanese industrial organization, distinctive societal characteristics, and cultural values.

Aspects of Japanese industrial organization have made it possible for MITI to intervene selectively without getting permanently entangled in the economy. Of the structural features that have facilitated selective government intervention, special mention should be made of career-long employment, the dual-structure economy (consisting of large parent companies and a vast network of small and medium-sized subcontractors and subsidiaries), pervasive intercorporate stockholding, and close banking-business ties. Such structural features have led to pronounced patterns of structural interdependence tying together organizations in Japan. The pervasiveness of structural interdependence is one of the most striking features of the Japanese political economy. It provides ready-made points of entry for selective government intervention.

Nor are ties of structural interdependence limited to the private sector. They also bind the private and public sectors together. The willingness of private industry to cooperate with MITI—even when company interests diverge from industrywide or national interests—resides in the long-term, "no-exit" relationship of interdependence binding corporations to MITI. Corporations depend on MITI for a variety of services, ranging from programs of technology push to mediation on matters of conflict resolution and consensus building. The greater the services rendered and received, all things being equal, the stronger the bonds of the relationship; the steel industry stands on the strong end of the spectrum and precision equipment producers on the weak end, with most high-technology industries located somewhere in between. For most industries, the bonding is stronger in Japan than in other advanced industrial countries.

The structure of interdependence is manifested in other ways. It is seen in the phenomenon of amakudari, the movement of higher civil servants from the various ministries into high-level, second-career posts in public corporations and the private sector. The movement of personnel creates cross-cutting linkages that help to consolidate the relationship of closeness and mutual trust. Public sector–private sector interdependence is also manifested in the myriad of informal, personal relationships between bureaucrats and business leaders. Such human networks (jinmyaku) function as channels of communication between the public and private sectors and as mechanisms for bridging the divide between "frame" organizations in Japan.[5] Higher civil servants in MITI utilize these personal contacts extensively in shaping industrial policy. Much of the crucial negotiation that goes into industrial policy-making takes place behind the scenes in what might be called an intermediate zone between the public and private sectors. It would be hard to formulate and implement Japanese industrial policy if the labyrinth of personal relationships in the intermediate zone did not exist.

What holds the structure of industrial organization together and helps to make the private sector competitive are deep-seated sociocultural values that fit in with Japan's system of capitalism. Certain Japanese values are already well known—the work ethic, frugality, honesty, strong meritocratic principles, and orientation toward collective goals; such values have been called the functional equivalent of the Protestant ethic.[6] But beyond the core values that meet the functional requisites of market capitalism, there are relational values that give capitalism in Japan a distinctive cast. Included here are the emphasis on the group over the individual; the stress on harmony; cooperation and competition; achievement and ascription; hierarchy and equity; obligation; long-term, "no-exit" commitments; reciprocity; the sharing of risks, costs, and benefits; and mutual trust. It is these values, the by-products of over a thousand years of social evolution, that give distinctive shape and life to the institutions of Japanese capitalism.[7]

Reciprocity, long-term commitments, obligation, mutual trust, and the sharing of risks, costs, and benefits, for example, help to hold together Japan's dual-structure economy; the relationship between large parent companies and their network of subcontractors and subsidiaries would be very different if Japanese buyers and suppliers did not adhere to this set of relational norms. The emphasis on collective goals also tempers the pursuit of parochial self-interest on the part of private corporations, and this fits in well not only with the structure of interdependence but with the capacity of government and business to form an identity of interests and reach consensus on a common set of collective goals.

Some social scientists, most notably economists, downplay the role of culture, emphasizing instead the importance of rational choice in explaining individual and group behavior. Other social scientists, especially anthropologists, assert the primacy of culture in providing a fundamental framework of meaning in terms of which all social action takes place. Certain political scientists and sociologists stress the importance of institutions in setting the parameters for political and economic behavior. The position taken in this book is that all three elements—rational choice, culture, and institutional structure—coexist in an interwoven relationship of complexity. Culture conditions rational choice and permeates institutional structure; rationality and institutional structures, in turn, give contextual shape to the ways in which individuals and groups draw upon and enact cultural values. The complexity of the interaction means that, contrary to economists who dismiss its significance, culture plays a central role in political-economic behavior. Its role in Japan's political economy is especially noteworthy not only because of differences in the configuration of values from those found in most Western capitalist countries, but also because of the way Japanese culture reinforces the maze of delicately meshed, interdependent institutions without leading to costly rigidities and economic inefficiencies.

MITI's capacity to implement a comparatively effective set of industrial policies for high technology has been due, in no small measure, to a combination of distinctive factors: the LDP's long dominance of the Diet; the relative weakness of labor-based, socialist parties; the light burden of military expenditures; Japan's large and homogeneous population; structural features of Japanese industrial organization; adaptable sociocultural endowments; and, of course, all the obvious factors to which attention has been drawn in this book, such as MITI's capacity to aggregate contending private-sector interests and the dynamism of the private sector. In short, the distinctiveness of Japanese industrial policy emerges from a complex configuration of factors, no one of which can possibly be singled out as decisive.

Reference Matter

Notes

Chapter One

1. In the United States, the "Japan, Incorporated" stereotype is widely held in the business community. See Richard D. Copaken, Andrew W. Singer, Oscar M. Garibald, and Michael P. Richman, *Petition to the President of the United States Through the Office of the United States Trade Representative for the Exercise of Presidential Discretion Authorized by Section 103 of the Revenue Act of 1971*, 26 U.S.C., 48 (a)(7)(D). Unpublished document issued by Houdaille Industries, Inc., May 3, 1982.

2. Adam Smith, *The Wealth of Nations*, edited by A. S. Skinner (London: Penguin Books, 1970).

3. Y. Ojimi, "Basic Philosophy and Objectives of Japanese Industrial Policy," in Organization for Economic Development (OECD), *The Industrial Policy of Japan* (Paris: OECD, 1972), pp. 11–31.

4. *Far Eastern Economic Review*, Aug. 20, 1987, p. 52.

5. Keizai Kōhō Center, *Japan: An International Comparison, 1986* (Tokyo: Keizai Kōhō Center, 1987), p. 83.

6. *Wall Street Journal*, Sept. 13, 1984, p. 26.

7. William G. Shepherd and Clair Wilcox, *Public Policies Toward Business* (Homewood, Ill.: Richard D. Irwin, 1979), pp. 405–74.

8. Daniel I. Okimoto, "Political Power in Japan," in Shumpei Kumon and Henry Rosovsky, eds., *The Political Economy of Japan*, vol. 3, *Culture* (Stanford, Calif.: Stanford University Press, forthcoming).

9. Eugene Kaplan, *Japan: The Government-Business Relationship* (Washington, D.C.: U.S. Department of Commerce, 1972).

10. Ira Magaziner and Thomas M. Hout, *Japanese Industrial Policy* (Berkeley: Institute of International Studies, University of California, 1980), pp. 73–74.

11. Victoria Curzon Price, *Industrial Policies in the European Community* (London: St. Martin's Press, 1984), pp. 17–83.

12. This is not to imply that MITI is incapable of orderly interest aggregation; with so many interests represented, overall coherence is hard to achieve.

13. Mancur Olson, *The Rise and Decline of Nations* (New Haven, Conn.: Yale University Press, 1982), pp. 36–74.

14. Daniel I. Okimoto, Henry Rowen, and Michael Dahl, *National Secu-*

rity and the Declining Competitiveness of the American Semiconductor Industry. Occasional Paper (Stanford, Calif.: Northeast Asia–United States Forum on International Policy, 1987).

15. Charles Schultze describes the deepening legal entanglements caused by the imposition of regulatory controls. High-powered corporate lawyers in the United States find hidden loopholes in the regulations; regulatory revisions are made to close the loopholes; lawyers search for new ways of circumventing the laws, and so on. The laws become very complex and the legal services needed to wend one's way through the legal labyrinth become costly. For both business and government, the transaction costs are formidable. Charles L. Schultze, *The Public Use of Private Interest* (Washington, D.C.: Brookings Institution, 1977).

16. Kōzō Yamamura, "Success That Soured: Administrative Guidance and Cartels," in Kōzō Yamamura, ed., *Policy and Trade Issues of the Japanese Economy* (Seattle: University of Washington Press, 1982), p. 96.

17. Ryūtarō Komiya, "Planning in Japan," in Morris Bornstein, ed., *Economic Planning: East and West* (Cambridge, Mass.: Ballinger, 1975), p. 221.

18. Frederick W. Richmond with Michael Kahan, *How to Beat the Japanese at Their Own Game* (Englewood Cliffs, N.J.: Prentice-Hall, 1983), pp. 101–4.

19. William V. Rapp, "Japan's Industrial Policy," in Isaiah Frank, ed., *The Japanese Economy in International Perspective* (Baltimore, Md.: Johns Hopkins University Press, 1975), p. 37.

20. Ibid. See also William V. Rapp, "Japan: Its Industrial Policies and Corporate Behavior," *Columbia Journal of World Business* 12, no. 1 (Spring 1977): 38–48.

21. Daniel I. Okimoto, *Pioneer and Pursuer: The Role of the State in the Evolution of the Japanese and American Semiconductor Industries*. Occasional Paper (Stanford, Calif.: Northeast Asia–United States Forum on International Policy, 1983), p. 56.

22. Ira Magaziner and Robert Reich, *Minding America's Business* (New York: Harcourt Brace Jovanovich, 1982), pp. 329–63.

23. Mancur Olson, *The Logic of Collective Action* (Cambridge Mass.: Harvard University Press, 1965).

24. Kenneth J. Arrow, "Economic Welfare and the Allocation of Resources for Invention," in National Bureau of Economic Research, ed., *The Rate and Direction of Inventive Activity* (Princeton, N.J.: Princeton University Press, 1962), pp. 609–25.

25. Arthur Gerstedfeld, ed., *Science Policy Perspectives: USA-Japan* (New York: Academic Press, 1982), pp. 103–206. See also Keith Jones, "Esprit de Corps in the Race for Technological Excellence," *Electronic News* 9, no. 8 (July 1983): 36–38.

26. U.S. government intervention, of course, is not solely reactive. Regulatory control over food and pharmaceuticals falls into the category of anticipatory action.

27. Eleanor M. Hadley, *Antitrust in Japan* (Princeton, N.J.: Princeton University Press, 1970).

28. Kōzō Yamamura, "Structure Is Behavior," in Isaiah Frank, ed., *The*

Japanese Economy in International Perspective (Baltimore, Md.: Johns Hopkins University Press, 1975), pp. 67–93.

29. Robert H. Bork, *The Antitrust Paradox: A Policy at War with Itself* (New York: Basic Books, 1978), pp. 15–106.

30. Lee Iacocca, "Lawsuits Make Risks Too Risky," *Japan Economic Journal*, Sept. 19, 1987, p. 7.

31. Peter Hall, *Governing the Economy: The Politics of State Intervention in Britain and France* (New York: Oxford University Press, 1986), pp. 34–47.

32. Ibid.

33. Gary R. Saxonhouse, "Japanese High Technology, Government Policy, and Evolving Comparative Advantage in Goods and Services," unpublished paper, 1982.

34. John Zysman, *Government, Markets, and Growth* (Ithaca, N.Y.: Cornell University Press, 1983), pp. 55–95.

35. Ronald P. Dore, "Goodwill and the Spirit of Market Capitalism," *British Journal of Sociology* 34, no. 4 (1983): 459–82.

36. Ronald P. Dore, *Flexible Rigidities: Industrial Policy and Structural Adjustment in the Japanese Economy 1970–80* (Stanford, Calif.: Stanford University Press, 1986), pp. 11–149.

37. Steven E. Rhoads, *The Economist's View of the World: Government, Markets, and Public Policy* (Cambridge, Eng.: Cambridge University Press, 1985), pp. 72–75.

38. Hugh Patrick, "Japanese Industrial Policy and Its Relevance for United States Industrial Policy," testimony before the Joint Economic Committee of the U.S. Congress, Washington, D.C., July 13, 1983.

39. Gary R. Saxonhouse, "International Economic and Technological Trends and the Changing World Order," unpublished paper prepared for the Japan Political Economy Research Conference, Honolulu, Hawaii, July 1983.

40. For an analysis of Japanese macroeconomic policies following the first and second oil crises, see Yōichi Shinkai, "Oil Crises and Stagflation in Japan," in Kōzō Yamamura, ed., *Policy and Trade Issues of the Japanese Economy* (Seattle: University of Washington Press, 1982), pp. 173–93.

41. Ryōkichi Hirono, *Factors Which Hinder or Help Productivity Improvement. Country Report: Japan* (Tokyo: Asian Productivity Organization, 1980), pp. 55–111.

42. Hiroya Ueno, "Industrial Policy: Its Role and Limits," *Journal of Japanese Trade and Industry* 2, no. 4 (July–Aug. 1983): 34–37.

43. William Lockwood, *The Economic Development of Japan* (Princeton, N.J.: Princeton University Press, 1968).

44. Myōhei Shinohara writes, "Whether it was steel, petrochemicals, or other industries, dissenting voices were raised claiming that the development of capital-intensive industries was irrational. The cost of international steel products was then comparatively high, and the industry was highly capital-intensive. In terms of classical comparative cost theory, such industries as textiles, apparel, and shipbuilding were in comparatively advantageous positons during the 1950's." *Industrial Growth, Trade, and Dynamic Patterns in the Japanese Economy* (Tokyo: University of Tokyo Press, 1982), p. 24.

45. By far the most thorough and stimulating historical treatment of Jap-

anese industrial policy is provided by Chalmers Johnson, *MITI and the Japanese Miracle: The Growth of Industrial Policy, 1925–1975* (Stanford, Calif.: Stanford University Press, 1982), pp. 198–274.

46. Komiya, pp. 189–235.

47. Tsunehiko Watanabe, "National Planning and Economic Development: A Critical Review of the Japanese Experience," *Economics of Planning* 10, no. 1–2 (1970): 21–51.

48. Compare planning in Japan with that in France. See Stephen S. Cohen, *Modern Capitalist Planning: The French Model* (Berkeley: University of California Press, 1977), especially parts I–IV.

49. Itō Mitsuharu, ed., *Sengo sangyō-shi e no shōgen*, vol. 1 (Hearings Concerning the Postwar History of Industry) (Tokyo: Mainichi Shimbunsha, 1977), pp. 13–57.

50. Ibid., pp. 159–79.

51. Ibid., pp. 183–277.

52. Terutomo Ozawa, *Japan's Technological Challenge to the West* (Cambridge, Mass.: MIT Press, 1974), pp. 16–30.

53. This assumes, of course, that greater market concentration lends itself to cooperation and consensus formation between government and the private sector—an assumption that may or may not be warranted, depending on the circumstances. And if big corporations in control of large market shares are not disposed to cooperate, the government may have a harder time imposing its will on such powerful and autonomous companies.

54. David Ricardo, *The Principles of Political Economy and Taxation* (New York: Penguin, 1971), ch. 7.

55. Funabashi Yōichi, *Keizai anzen hoshō-ron* (An essay on economic security) (Tokyo: Tōyō Keizai Shimpōsha, 1978), pp. 20–148. Ministry of International Trade and Industry, *Economic Security of Japan* (Tokyo: MITI, 1982).

56. Yoshi Tsurumi, *The Japanese Are Coming* (Cambridge, Mass.: Ballinger, 1976), pp. 37–69.

57. The petrochemical complex built in Iran during the late 1970's is an example of a government-backed overseas project, even though it started as a purely private-sector endeavor. MITI turned it into a national project after the Shah of Iran was overthrown in 1979, and the plant facilities sustained damages during the Iran-Iraq hostilities.

58. Amaya Naohiro, *Nihon kabushiki kaisha: Nokosareta sentaku* ("Japan, Incorporated": The remaining options) (Tokyo: PHP, 1982), pp. 57–99.

59. Daniel I. Okimoto, "Arms Transfers: The Japanese Calculus," in John H. Barton and Ryukichi Imai, eds., *Arms Control II* (Cambridge, Mass.: Oelgeschlager, Gunn & Hain, 1981), pp. 273–317.

60. The Japanese are worried about the complications of national security considerations in the United States with respect to foreign imports, technology transfers, and direct foreign investments.

61. Peter J. Katzenstein, *Small States in World Markets: Industrial Policy in Europe* (Ithaca, N.Y.: Cornell University Press, 1985).

62. Peter J. Katzenstein, "Japan, Switzerland of the Far East?" in Takashi Inoguchi and Daniel I. Okimoto, eds., *The Political Economy of Japan*, vol. 2,

The Changing International Context (Stanford, Calif.: Stanford University Press, 1988).

63. Komiya, p. 221.

64. Interviews with over 50 leaders from government, business, labor, academia, and the mass media.

65. Yamamura, "Success That Soured," pp. 77–112.

66. Amaya Naohiro, "Wa no rinri to dokkinhō no ronri" (The ethics of harmony and the logic of the anti-monopoly law), *Bungei shunjū* 58, no. 12 (Dec. 1980): 176–93.

67. Martin Bronfenbrenner, "Excessive Competition in Japanese Business," *Monumenta Nipponica* 21 (1966): 114–24.

68. Forward pricing refers to the strategy of setting prices close to, or sometimes even below, costs in the expectation that the sacrifice in short-term profits will be more than made up by the increase in demand (resulting from undercutting prices charged by competitors) and the consequent drop in the per-unit costs of production. It can be the same as predatory pricing, depending on the levels at which prices are set. Predatory pricing sets prices below costs in order to weaken or destroy competition.

69. Daniel I. Okimoto, Takuo Sugano, and Franklin B. Weinstein, eds., *Competitive Edge: The Semiconductor Industry in the U.S. and Japan* (Stanford, Calif.: Stanford University Press, 1984), pp. 179–213.

70. Amaya Naohiro, "Dokkinhō kaisei shian ni hanron suru" (Rebutting the proposed revision of the antimonopoly act), *Ekonomisuto*, Nov. 14, 1974, pp. 36–45.

71. Yamamura, "Success That Soured," pp. 77–112.

72. Amaya, p. 40.

73. The implicit view is that MITI did not seek to cultivate excessive competition but that it emerged from economic conditions that industrial policy inadvertently created.

74. Johnson, pp. 206–7.

75. Peter F. Drucker, "Economic Realities and Enterprise Strategy," in Ezra F. Vogel, ed., *Modern Japanese Organization and Decision-making* (Berkeley: University of California Press, 1975), pp. 228–44.

76. Kagono Tadao, Nonaka Ikujiro, Sakakibara Kiyonori, and Okumura Akihiro, *Nichibei kigyō no keiei hikaku* (Management of Japanese and American firms: a comparison) (Tokyo: Nihon Keizai Shimbunsha, 1983), pp. 24–25.

77. Drucker, pp. 239–43. Drucker argues that the market share approach of highly leveraged Japanese firms and the profit maximization approach of equity-financed U.S. companies are both rational responses to differences in the structure of capital markets. Both have the same objectives: "minimizing the cost of capital and minimizing the risk of not being able to obtain capital. These two objectives are the only rational foundation for a valid 'theory of the firm'" (p. 235).

78. Okumura Hiroshi, "Masatsu o umu Nihonteki keiei no heisasei" (Frictions that arise from the closed nature of Japanese management), *Ekonomisuto*, July 6, 1982, pp. 34–40.

79. George Stalk and Kenneth Arbour, "You Can Stop Pitying the Japanese Stockholder," *Wall Street Journal*, June 11, 1984.

80. John S. R. Shad, "The Leveraging of America," *Wall Street Journal*, June 8, 1984.

81. Masahiko Aoki, "The Japanese Firm in Transition," in Kōzō Yamamura and Yasukichi Yasuba, eds., *The Political Economy of Japan*, vol. 1, *The Domestic Transformation* (Stanford, Calif.: Stanford University Press, 1987), p. 287.

82. Yasusuke Murakami and Kōzō Yamamura, "A Technical Note on Japanese Firm Behavior and Economic Policy," in Kōzō Yamamura, ed., *Policy and Trade Issues of the Japanese Economy* (Seattle: University of Washington Press, 1982), pp. 113–21.

83. Ibid.

84. Shumpei Kumon, Yasusuke Murakami, and Kōzō Yamamura, "To Leave a Bamboo Thicket," unpublished paper presented at the Japan Political Economy Research Conference, Honolulu, Hawaii, July 1983, p. 65.

85. MITI officials point out that foreign objections to the organization of antirecession cartels on the grounds that they must go hand in hand with import restrictions are no longer valid. Although antirecession cartels or rationalization cartels have been formed in several of the basic materials industries that have been hard-hit by the recession and high costs of energy, the percentage of foreign imports has increased significantly, indicating that Japan's market has not been closed (see following table).

Industry Imports as a Share of the Japanese Market

Industry	1978	1979	1980	1981	1982
Petrochemicals	3.7%	4.0%	5.7%	8.7%	10.1%
Aluminum	43.0	39.0	45.0	61.0	63.0
Ferrosilicon	30.0	27.0	32.0	48.0	61.0

SOURCE: MITI, "Answers to Some of the Most Common Charges Against Japanese Industrial Policy," unpublished paper, Tokyo, May 1983, p. 39.

86. Ojimi, p. 27.

87. Assar Lindbeck, "Can Pluralism Survive?" William K. McInally Lectures (Ann Arbor: Graduate School of Business Administration, University of Michigan, 1977), p. 14.

88. Okimoto, Sugano, and Weinstein, pp. 78–133.

Chapter Two

1. Eugene Kaplan, *Japan: The Government-Business Relationship* (Washington, D.C.: U.S. Department of Commerce, 1972).

2. Daniel I. Okimoto, Henry Rowen, and Michael Dahl, *National Security and the Declining Competitiveness of the American Semiconductor Industry*. Occasional Paper (Stanford, Calif.: Northeast Asia–United States Forum on International Policy, 1987).

3. Semiconductor Industry Association, *The Effect of Government Targeting on World Semiconductor Competition* (Cupertino, Calif.: Semiconductor Industry Association, 1983).

4. Michael Borrus, James Milstein, and John Zysman, *U.S.-Japanese Competition in the Semiconductor Industry* (Berkeley: Institute of International Studies, University of California, 1982).

5. Daniel I. Okimoto, "Political Context," in Daniel I. Okimoto, Takuo Sugano, and Franklin B. Weinstein, eds., *Competitive Edge: The Semiconductor Industry in the U.S. and Japan* (Stanford, Calif.: Stanford University Press, 1984), pp. 78–113.

6. Mancur Olson, *The Logic of Collective Action* (Cambridge, Mass.: Harvard University Press, 1965).

7. Okimoto, Rowen, and Dahl.

8. Masahiko Aoki, "The Japanese Firm in Transition," in Kōzō Yamamura and Yasukichi Yasuba, eds., *The Political Economy of Japan*, vol. 1, *The Domestic Transformation* (Stanford, Calif.: Stanford University Press, 1987), pp. 282–86. See Aoki's article "Innovative Adaptation Through the Quasi-Tree Structure: An Emerging Aspect of Japanese Entrepreneurship," *Zeitschrift für Nationalökonomie*, suppl. 4 (1984): 177–98.

9. Semiconductor Industry Association.

10. United States International Trade Commission, *Foreign Industrial Targeting and Its Effects on U.S. Industries, Phase I: Japan* (Washington, D.C.: U.S. Government Printing Office, 1983), p. 17.

11. Nakagawa Yasuzō, *Hihon no handōtai kaihatsu* (The development of semiconductors in Japan) (Tokyo: Daiyamondosha, 1981), pp. 57–60.

12. Daniel I. Okimoto and Gary R. Saxonhouse, "Technology and the Future of the Economy," in Kōzō Yamamura and Yasukichi Yasuba, eds., *The Political Economy of Japan*, vol. 1, *The Domestic Transformation* (Stanford, Calif.: Stanford University Press, 1987), pp. 385–419.

13. Daniel I. Okimoto, *Pioneer and Pursuer: The Role of the State in the Evolution of the Japanese and American Semiconductor Industries*. Occasional Paper (Stanford, Calif.: Northeast Asia–United States Forum on International Policy, 1983), pp. 3–20.

14. Daniel I. Okimoto and Henry K. Hayase, "Organizing for Innovation," unpublished paper prepared for High Technology Conference, Honolulu, Hawaii, Jan. 1985.

15. United States International Trade Commission, p. 77.

16. Semiconductor Industry Association.

17. Makoto Kikuchi, *Japanese Electronics* (Tokyo: Simul Press, 1983), p. 6.

18. Interview with MITI official, Jan. 1984.

19. Daniel I. Okimoto, Takuo Sugano, and Franklin B. Weinstein, eds., *Competitive Edge: The Semiconductor Industry in the U.S. and Japan* (Stanford, Calif.: Stanford University Press, 1984), pp. 38–39.

20. Masato Nebashi, "VLSI Technology Research Association," unpublished paper, 1981; interview with Masato Nebashi, June 24, 1982.

21. Edward A. Feigenbaum and Pamela McCorduck, *The Fifth Generation: Artificial Intelligence and Japan's Computer Challenge to the World* (Menlo Park, Calif.: Addison-Wesley, 1983).

22. Nakagawa, pp. 22–32.

23. United States International Trade Commission, p. 111.

24. Arthur D. Little, "Summary of Major Projects in Japan for R&D of Information Processing Technology," unpublished study, 1983.

25. Interview with Daimaru executives, Dec. 1983 and Feb. 1987.

26. Oliver E. Williamson, *Markets and Hierarchies: Analysis and Antitrust Implications* (New York: Free Press, 1975).

27. Masahiko Aoki, *The Cooperative Game Theory of the Firm* (Oxford: Oxford University Press, 1984).

28. Ronald P. Dore, *Flexible Rigidities: Industrial Policy and Structural Adjustment in the Japanese Economy, 1970–80* (Stanford, Calif.: Stanford University Press, 1986).

29. Okimoto and Hayase.

30. Charles L. Schultze, *The Public Use of Private Interest* (Washington, D.C.: Brookings Institution, 1977).

31. United States International Trade Commission, p. 90.

32. Semiconductor Industry Association, pp. 99–101.

33. Okimoto, *Pioneer and Pursurer*, p. 4.

34. Williamson, pp. 26–31.

35. United States International Trade Commission, p. 105.

36. Interviews with MITI officials, Sept. 1983.

37. Feigenbaum and McCorduck, p. 109.

38. Ibid.

39. "Bell Labs: The Threatened Star of U.S. Research," *Business Week,* July 5, 1982, p. 47.

40. "A Research Spending Surge Defies Recession," *Business Week,* July 5, 1982, p. 54.

41. Glenn R. Fong, "Industrial Policy Innovation in the United States: Lessons from the Very High Speed Integrated Circuit Program," unpublished paper presented at the annual meeting of the American Political Science Association, Chicago, Sept. 1–4, 1983.

42. National Science Foundation, "Industrial R&D Expenditures in 1980 Show Real Growth for Fifth Consecutive Year," *Highlights,* Dec. 31, 1981, pp. 81–331.

43. Robert W. Wilson, Peter K. Ashton, and Thomas P. Egan, *Innovation, Competition, and Government Policy in the Semiconductor Industry* (Lexington, Mass.: D.C. Heath, 1980), p. 154.

44. Nico Hazewindus with John Tooker, *The U.S. Microelectronics Industry* (New York: Pergamon Press, 1982), p. 121.

45. Keith Jones, "French and U.S. Interests Intertwine Around ICs," *Electronic Business* 8, no. 7 (1983): 88.

46. Keith Jones, "Nationalized French Computer Industry Sparks Controversy," *Electronic Business* 9, no. 4 (1983): 37–38.

47. "Europe's Desperate Try for High-Tech Teamwork," *Business Week,* May 30, 1983, p. 45.

48. United States International Trade Commission.

49. Ibid.

50. Thomas P. Rohlen, "Learning: The Mobilization of Knowledge in the Japanese Political Economy," unpublished paper prepared for the Japan Political Economy Research Conference, Tokyo, Jan. 6–10, 1988.

51. Robert Michels, *Political Parties: A Sociological Study of the Oligarchical Tendencies of Modern Democracy* (New York: Dover, 1959).

52. Gary R. Saxonhouse, "Japanese High Technology, Government Policy, and Evolving Comparative Advantage in Goods and Services," unpublished paper, 1982.

53. Ibid.

54. United States International Trade Commission, p. 76.

55. Fong.

56. Interviews with Silicon Valley executives, Sept. 1982, Apr. 1986, June 1987.

57. Interview with Masato Nebashi, former executive director, VLSI project, Tokyo, June 24, 1982.

58. Nihon Denshi Keisanki Kabushiki Kaisha, *JECC Computer Handbook* (Tokyo: Nihon Denshi Keisanki Kabushiki Kaisha, 1983).

59. Chalmers Johnson, *MITI and the Japanese Miracle: The Growth of Industrial Policy, 1925–1975* (Stanford, Calif.: Stanford University Press, 1982).

60. Sheridan Tatsuno, *The Technopolis Strategy: Japan, High Technology, and the Control of The 21st Century* (Englewood Cliffs, N.J.: Prentice Hall, 1986).

61. Tanaka Kakuei, *Nihon rettō kaizō-ron* (A plan for restructuring Japan) (Tokyo: Nikkan Kōgyō Shinbun, 1972).

62. Morton I. Kamien and Nancy L. Schwartz, *Market Structure and Innovation* (Cambridge, Eng.: Cambridge University Press, 1982), pp. 35–48.

63. Richard Levin, "The Semiconductor Industry," in Richard R. Nelson, ed., *Government and Technical Progress* (New York: Pergamon Press, 1982), pp. 9–100.

64. Interview with Elliott Levinthal, formerly of the Defense Advanced Research Project Agency, U.S. Department of Defense, Sept. 1983.

65. Richard R. Nelson, "Government Stimulus of Technological Progress: Lessons from American History," in Richard R. Nelson, ed., *Government and Technical Progress* (New York: Pergamon Press), pp. 471–72.

66. Robert L. Muller, "U.S. Computer Firms in Britain Suspect U.K. Agencies Favor Home Companies," *Wall Street Journal*, Aug. 12, 1982.

67. Nihon Jōhō Shori Kaihatsu Kyōkai, *Conpyuuta hakusho, 1981* (Computer white paper) (Tokyo: Nihon Jōhō Shori Kaihatsu Kyōkai, 1981), pp. 95–105.

68. Yasusada Kitahara, *Information Network System: Telecommunications in the Twenty-first Century* (London: Heinemann Educational Books, 1983), pp. 80–103.

69. Okimoto and Hayase.

70. Nihon Denshi Keisanki Kabushiki Kaisha.

71. Daniel I. Okimoto, "Outsider Trading: Coping with Japanese Industrial Organization," *Journal of Japanese Studies* 13, no. 2 (1987): 85–116.

72. Williamson.

73. Hiroshi Okumura, "The Closed Nature of Japanese Intercorporate Relations," *Japan Echo* 9, no. 3 (1982): 61.

74. Williamson; Harvey Liebenstein, "The Japanese Management System: An X-Efficiency-Game Theory Analysis," in Masahiko Aoki, *The Economic*

Analysis of the Japanese Firm (Amsterdam: North-Holland, 1984), pp. 331–53.

75. Interviews with Silicon Valley executives, 1983 to 1987.

76. Interview with Art Hausmann, former chairman of Ampex Corporation, June 1984.

77. Gary R. Saxonhouse, "Industrial Policy and Factor Markets: Biotechnology in Japan and the United States," in Hugh T. Patrick, ed., *Japanese High Technology Industries* (Seattle: University of Washington Press, 1986).

78. Peter J. Katzenstein, *Small States in World Markets: Industrial Policy in Europe* (Ithaca, N.Y.: Cornell University Press, 1985).

Chapter Three

1. The analysis will be brief, because Chalmers Johnson has written a comprehensive and insightful book on MITI, *MITI and the Japanese Miracle: The Growth of Industrial Policy, 1925–1975* (Stanford, Calif.: Stanford University Press, 1982).

2. The most compelling case is made by Johnson in *MITI and the Japanese Miracle.*

3. Ibid., pp. 49–50.

4. John A. Armstrong, *The European Administrative Elite* (Princeton, N.J.: Princeton University Press, 1973), pp. 73–199; Ezra N. Suleiman, *Elites in French Society* (Princeton, N.J.: Princeton University Press, 1978), pp. 17–92.

5. For a relatively optimistic assessment of the possibilities for a coherent U.S. industrial policy, see Michael L. Wachter and Susan M. Wachter, eds., *Toward a New U.S. Industrial Policy?* (Philadelphia: University of Pennsylvania Press, 1981).

6. I am grateful to the following MITI officials for their insights into MITI's internal organization: Isayama Takeshi, Katō Fumihiko, Funaki Takashi, Inaba Kenji, Shiozawa Bunro.

7. On the high costs of underwriting economic inefficiency in Europe, see Victoria Curzon Price, *Industrial Policies in the European Community* (London: St. Martin's Press, 1984), pp. 84–118.

8. Japan Institute of Labour, *Japanese Working Life Profile, 1986* (Tokyo: Japan Institute of Labour, 1986), pp. 56–57.

9. Ronald P. Dore, *Taking Japan Seriously: A Confucian Perspective on Leading Economic Issues* (Stanford, Calif.: Stanford University Press, 1987), pp. 68–84.

10. For a discussion of how the wage bargaining process works, see Dore, *Taking Japan Seriously,* pp. 70–73.

11. Rodney Clark, *The Japanese Company* (New Haven, Conn.: Yale University Press, 1979), pp. 55–64.

12. Robert H. Hayes and William J. Abernathy, "Managing Our Way to Economic Decline," *Harvard Business Review,* July-Aug. 1980, pp. 75–76; Robert B. Reich, *The Next American Frontier: A Provocative Program for Economic Renewal* (New York: Times Book, 1983), pp. 145–60.

13. The disposition to form independent subsidiaries is perhaps one of the reasons the so-called dual structure persists in Japan's postwar economy.

14. Gary R. Saxonhouse, "Industrial Policy and Factor Markets: Biotechnology in Japan and the United States," in Hugh T. Patrick, ed., *Japanese High Technology Industries* (Seattle: University of Washington Press, 1986), pp. 97–135.

15. Chūshō Kigyō-chō, ed., *Zu de miru chūshō kigyō hakusho* (The small and medium-sized enterprise white paper: Tables and graphs) (Tokyo: Dōyukan, 1980), p. 61.

16. Ken'ichi Imai, "Japan's Changing Industrial Structure and United States–Japan Industrial Relations," in Kōzō Yamamura, ed., *Policy and Trade Issues of the Japanese Economy* (Seattle: University of Washington Press, 1982), p. 64.

17. Richard E. Caves and Masu Uekusa, *Industrial Organization in Japan* (Washington, D.C.: Brookings Institution, 1976), p. 113.

18. Gary R. Saxonhouse, "Industrial Restructuring in Japan," *Journal of Japanese Studies* 5, no. 2 (Summer 1979): 289–95.

19. Miyohei Shinohara, "A Survey of the Japanese Literature on Small Industry," in Bert F. Hoselitz, ed., *The Role of Small Industry in the Process of Economic Growth* (New York: Humanities Press, 1968), pp. 1–113.

20. Masahiko Aoki, "A Few Facts on the Ownership and Equity Structure of Large Japanese Firms," unpublished paper presented to the U.S.-Japan High Technology Research Project, Northeast Asia–United States Forum on International Policy, Stanford University, Stanford, Calif., Nov. 1982, p. 30.

21. Hirschman points out that when there is no barrier to exit, the tendency to exit "deprives the faltering firm or organization of those who could best help it fight its shortcomings and its difficulties." Albert O. Hirschman, *Exit, Voice, and Loyalty: Responses to Decline in Firms, Organizations, and States* (Cambridge, Mass.: Harvard University Press, 1970), p. 79.

22. However, even with this multilayered system of safety nets, a very large number of small and medium-sized companies still fall by the wayside, particularly in sectors where Japan is losing comparative advantage.

23. Ken'ichi Imai and Hiroyuki Itami, "Mutual Infiltration of Organization and Market: Japan's Firm and Market in Comparison with the U.S." (Tokyo: Institute of Business Research, Hitotsubashi University, Feb. 1984). See also Imai Ken'ichi, *Nihon no sangyō shakai* (Japan's industrial society) (Tokyo: Chikuma Shoten, 1983), pp. 103–8.

24. Oliver E. Williamson, *Markets and Hierarchies: Analysis and Antitrust Implications* (New York: Free Press, 1975).

25. On Japan's six major keiretsu groups, see Okumura Hiroshi, *Nihon no roku daikigyō shūdan* (Japan's six giant industrial groups) (Tokyo: Daiyamondo, 1976), pp. 170–243.

26. Eleanor M. Hadley, *Antitrust in Japan* (Princeton, N.J.: Princeton University Press, 1970), pp. 205–56.

27. Masahiko Aoki, "Innovative Adaptation Through the Quasi-tree Structure: An Emerging Aspect of Japanese Entrepreneurship," *Zeitschrift für Nationalökonomie*, suppl. 4 (1984): 177–98.

28. M. Therese Flaherty and Hiroyuki Itami, "Finance," in Daniel I. Okimoto, Takuo Sugano, and Franklin B. Weinstein, eds., *Competitive Edge: The Semiconductor Industry in the U.S. and Japan* (Stanford, Calif.: Stanford University Press, 1984), p. 146.

29. Ibid., p. 151.

30. Hirschman, pp. 120–26.

31. Daniel I. Okimoto, "Outsider Trading: Coping with Japanese Industrial Organization," *Journal of Japanese Studies* 13, no. 2 (1987): 389.

32. Okumura Hiroshi, "Masatsu o umu Nihonteki keiei no heisasei" (Frictions that arise from the closed nature of Japanese management), *Ekonomisuto*, July 6, 1982, pp. 34–40.

33. Iwao Nakatani, "The Economic Role of Financial Corporate Grouping in Japan," in Masahiko Aoki, *The Economic Analysis of the Japanese Firm* (Amsterdam: North-Holland, 1984), pp. 227–58.

34. This presumes, of course, that there is a joint utility function for all the relevant corporate constituents and that it can be specified.

35. Richard Pascale and Thomas P. Rohlen, "The Mazda Turnaround," in Daniel I. Okimoto and Thomas P. Rohlen, eds., *Inside the Japanese System: Readings on Contemporary Society and Political Economy* (Stanford, Calif.: Stanford University Press, 1988), pp. 149–69.

36. Caves and Uekusa, pp. 47–58.

37. Okumura Hiroshi, *Hōjin shihonshugi no kōzō* (The structure of corporate capitalism) (Tokyo: Nihon Hyōronsha, 1975), pp. 81–110.

38. Aoki, "Innovative Adaptation," p. 8.

39. John Zysman, *Government, Markets, and Growth* (Ithaca, N.Y.: Cornell University Press, 1983), pp. 55–95.

40. Ibid., p. 63.

41. Sakakibara Eisuke and Noguchi Yukio, "Ōkurashō-Nichigin ōchō no bunseki" (An analysis of the almighty power of the Ministry of Finance and the Bank of Japan), *Chūō kōron*, Aug. 1977, pp. 96–150.

42. Flaherty and Itami, p. 145.

43. Hirschman, p. 96.

44. Stephen D. Krasner, "U.S. Commercial and Monetary Policy: Unravelling the Paradox of External Strength and Internal Weakness," *International Organization* 31, no. 4 (Autumn 1977): 644.

45. Ibid.

46. Caves and Uekusa, p. 27.

47. The past history of MITI interaction with such basic, old-line industries as steel and petroleum refining has provided the grist for foreign perceptions of "Japan, Inc." and the pervasive notion of "controlled competition." Although the two cases substantiate such perceptions, the experiences of other industries do not. Certainly the information industries cannot be fitted into such stereotypical molds.

48. Ken'ichi Imai, "Interfirm Group Behaviors in Japanese Industrial Organization," unpublished paper, Feb. 16, 1979, pp. 11–14.

49. Peter J. Katzenstein, "Conclusion: Domestic Structures and Strategies," *International Organization* 31, no. 4 (Autumn 1977): 892.

50. Imai Ken'ichi, Itami Hiroyuki, and Koike Kazuo, *Naibu soshiki no keizaigaku* (The economics of internal organizations) (Tokyo: Tōyō Keizai Shimpōsha, 1982). Chalmers Johnson, *Japan's Public Policy Companies* (Washington, D.C.: American Enterprise Institute, 1978).

51. The categories are based on Chalmers Johnson's typology of public corporations. Johnson, *Japan's Public Policy Companies*, p. 27.

52. Imai Ken'ichi, Itami Hiroyuki, and Koike Kazuo, pp. 131–34.

53. For a handy chronological listing of Japan's 112 public corporations, see the appendix in Johnson, *Japan's Public Policy Companies,* pp. 149–66.

54. Williamson, pp. 26–28.

55. Interview conducted with former deputy directors of the Electronics Policy, Data Processing Promotion, and Electronics and Electrical Machinery divisions from 1981 to 1984.

56. Charles L. Schultze, *The Public Use of Private Interest* (Washington, D.C.: Brookings Institution, 1977), pp. 5–15.

57. Suzanne D. Berger, ed., *Organizing Interests in Western Europe: Pluralism, Corporatism, and the Transformation of Politics* (Cambridge, Eng.: Cambridge University Press, 1981). See especially the chapters by Claus Offe, Alessandro Pizzorno, Philippe C. Schmitter, and Juan Linz.

58. Gordon Adams, *The Iron Triangle* (New York: Council on Economic Priorities, 1981), pp. 84–85.

59. C. Andrea Bollino, "Industrial Policy in Italy: A Survey," and Francois DeWitt, "French Industrial Policy from 1945–1981: An Assessment," in F. Gerard Adams and Lawrence R. Klein, eds., *Industrial Policies for Growth and Competitiveness* (Lexington, Mass.: D.C. Heath, 1983), pp. 263–304 and pp. 221–46.

60. Jinji-in, *Eiri kigyō e no shūshoku no shōnin ni kansuru nenji hokokushō* (Annual report on approval for employment in private enterprise) (Tokyo: Jinji-in, 1976).

61. Adams, pp. 84–85.

62. Chalmers Johnson, "The Reemployment of Retired Government Bureaucrats in Japanese Big Business," *Asian Survey* 14 (Nov. 1974): 953–65; see also Johnson's discussion in *Japan's Public Policy Companies,* pp. 105–14.

63. This criticism has been voiced by Japanese journalists and academics as well as by foreign businessmen who see amakudari as an integral part of "Japan, Inc."

64. Ouchi Minoru, *Fuhai no kōzō: Ajiateki kenryoku no tokushitsu* (Structural corruption: Characteristics of Asian political power) (Tokyo: Daiyamondo-sha, 1977), pp. 193–96.

65. This estimate is based on unsystematic "head counts" made by MITI officials from the Minister's Secretariat, not on a thorough examination of personnel records. It may therefore underestimate the actual number.

66. In the United States, Schultze writes, "instead of creating incentives so that public goals become private interests, private interests are left unchanged and obedience to the public is commanded." Hence the labyrinth of regulatory controls and litigation. Schultze, p. 6.

67. Textiles is a prime example of U.S. pressures and domestic resistance; see the excellent study by I. M. Destler, Haruhiro Fukui, and Hideo Sato, *The Textile Wrangle* (Ithaca, N.Y.: Cornell University Press, 1978).

68. Chie Nakane, *Japanese Society* (Berkeley: University of California Press, 1970), pp. 87–103.

69. Ibid., p. 102.

70. Ibid.

71. Chitoshi Yanaga, *Big Business in Japanese Politics* (New Haven, Conn.: Yale University Press, 1968), pp. 32–62.

Chapter Four

1. Japan, Sōrifu, *Kokumin seikatsu chōsa* (Public opinion on people's lifestyles) (Tokyo: Ōkurashō Insatsukyoku, 1983).

2. NHK Yoron Chōsajo, *Daini Nihonjin no ishiki: NHK yoron chōsa* (Japanese attitudes: NHK opinion survey) (Tokyo: Shiseidō, 1980), pp. 306–9.

3. Yasusuke Murakami, "Toward a Socioinstitutional Explanation of Japan's Economic Performance," in Kōzō Yamamura, ed., *Policy and Trade Issues of the Japanese Economy* (Seattle: University of Washington Press, 1982) pp. 28–31.

4. Walter Galenson and Konosuke Odaka, "The Japanese Labor Market," in Hugh T. Patrick and Henry Rosovsky, eds., *Asia's New Giant* (Washington, D.C.: Brookings Institution, 1975), pp. 600–607.

5. Daniel I. Okimoto, *Pioneer and Pursuer: The Role of the State in the Evolution of the Japanese and American Semiconductor Industries*. Occasional Paper (Stanford, Calif.: Northeast Asia–United States Forum on International Policy, 1983), pp. 52–54.

6. Inoguchi Takashi, *Gendai Nihon seiji keizai no kozu* (The design of politics and economics in contemporary Japan) (Tokyo: Tōyō Keizai Shinpōsha, 1983), pp. 18–22.

7. Joji Watanuki, *Politics in Postwar Japanese Society* (Tokyo: University of Tokyo Press, 1977), p. 94.

8. Murakami, pp. 28–31.

9. Aurelia George, "The Japanese Farm Lobby and Agricultural Policy Making," *Pacific Affairs* 4, no. 3 (Fall 1981): 409–30.

10. Seiji Kōhō Sentaa, *Seiji handobukku, 1982* (Political handbook, 1982) (Tokyo, 1982), pp. 244, 263.

11. Ishikawa Masumi, *Sengo seiji kōzō-shi* (A history of the postwar structure of politics) (Tokyo: Nihon Hyōronsha, 1978), pp. 85–98.

12. Ibid.

13. Kenzo Hemmi, "Agriculture and Politics in Japan," in Emery N. Castle and Kenzo Hemmi, with Sally A. Skillings, *U.S.-Japanese Agricultural Trade Relations* (Washington, D.C.: Resources for the Future, 1982), p. 224.

14. Inoguchi, pp. 199–252.

15. The LDP's two-pronged strategy implies that LDP leaders based their policies on the assumption that Japanese voters act on their retrospective evaluation of the government's performance in office. This fits in with the retrospective and "what have you done for me lately" schools of election behavior. See Morris P. Fiorina, *Retrospective Voting in American National Elections* (New Haven, Conn.: Yale University Press, 1981), and Samuel Popkin, W. Gorman, C. Phillips, and J. A. Smith, "Comment: What Have You Done for Me Lately? Toward an Investment Theory of Voting," *American Political Science Review* 70, no. 3 (Sept. 1976): 779–805.

16. Kazuo Satō, "Japan's Savings and Internal and External Macroeconomic Balance," in Kōzō Yamamura, ed., *Policy and Trade Issues of the Japanese Economy* (Seattle: University of Washington Press, 1982), pp. 143–72.

17. Saitō Jirō et al., *Kuni no yosan, 1984* (The national budget, 1984) (Tokyo: Hase Shobō, 1984), chap. 2.

18. The JCP sought to galvanize support within the small and medium-sized sectors by organizing a group called Minshu Shōkōkai, or *Minshō*, the People's Chamber of Commerce, which the Japan Chamber of Commerce, Nisshō, considered a potential threat to LDP dominance. See Shirai Hisaya, *Kiki no naka no zaikai* (Big business faced with crisis) (Tokyo: Saimaru Shuppankai, 1973), pp. 9–23.

19. Ibid., p. 23.

20. Hemmi, pp. 252–53.

21. Inoguchi, pp. 224–30.

22. Yoichi Shinkai, "Oil Crises and Stagflation in Japan," in Kōzō Yamamura, ed., *Policy and Trade Issues of the Japanese Economy* (Seattle: University of Washington Press, 1982), pp. 173–87.

23. The lack of a viable alternative party or coalition can inhibit the predictive power of the theory of retrospective voting; but the voters' retrospective judgments can also be conditioned by inferences about hypothetical alternatives. See James E. Alt and K. Alec Chrystal, *Political Economics* (Berkeley: University of California Press, 1983), pp. 156–59.

24. NHK, Nihon Hōsō Kyōkai (NHK) Hōsō Yoron Chōsajo (Japan Broadcasting Corporation Public Opinion Survey Center), ed., *Daini Nihonji no ishiki: NHK yoron chōsa* (Japanese Attitudes: NHK Public Opinion Survey) (Tokyo: Shiseidō, 1980) pp. 306–9.

25. Yasusuke Murakami, "The Age of New Middle Mass Politics: The Case of Japan," *Journal of Japanese Studies* 8, no. 1 (Winter 1982): 64.

26. Gerard Kramer, "Electoral Politics in the Zero-sum Society," paper presented at annual meeting of the American Political Science Association, Sept. 1–5, 1982, Denver, Colorado. See also Popkin, Gorman, Phillips, and Smith, pp. 779–805; Fiorina; Douglas Hibbs, "Economic Outcomes and Political Support for British Governments Among Occupational Classes: A Dynamic Analysis," *American Political Science Review* 76 (1982): 259–79.

27. Yasusuke Murakami, "Age of New Middle Mass," pp. 69–72.

28. The classic work on power elites is C. Wright Mills, *The Power Elite* (New York: Oxford University Press, 1959). See also the seminal work by Robert Michels, *Political Parties: A Sociological Study of the Oligarchical Tendencies of Modern Democracy* (New York: Dover, 1959).

29. The classic work in the liberal pluralist literature is Robert A. Dahl, *Who Governs? Democracy and Power in an American City* (New Haven, Conn.: Yale University Press, 1961).

30. On corporatist theory, see Suzanne D. Berger, ed., *Organizing Interests in Western Europe: Pluralism, Corporatism, and the Transformation of Politics* (Cambridge, Eng.: Cambridge University Press, 1981), especially the chapter by Philippe C. Schmitter, pp. 287–330. For an application of corporatist theory to Japan, see T. J. Pempel and Keiichi Tsunekawa, "Corporatism Without Labor? The Japanese Anomaly," in Philippe Schmitter and Gerhard Lehmbruch, eds., *Trends Toward Corporatist Intermediation* (Beverly Hills, Calif.: Sage, 1979), pp. 231–70.

31. Chitoshi Yanaga portrays Keidanren as more powerful than it is. See

Big Business in Japanese Politics (New Haven, Conn.: Yale University Press, 1968), pp. 68–72.

32. One of the best examples of the statist literature is Stephen D. Krasner, *Defending the National Interest: Raw Materials Investments and U.S. Foreign Policy* (Princeton, N.J.: Princeton University Press, 1978).

33. Hirose Michisada, "'Rieki haibun shisutemu' wa henka shita ka?" (Has the system of distributing profits changed?), *Sekai,* Mar. 1983, p. 105.

34. For an inside account of the role of a construction company in Shiga Prefecture, see Tonooka Akio, "Seijika ni okeru kane no kenkyū" (Research on money in the hands of politicians), *Bungei Shunjū,* June 1975, pp. 160–94.

35. Hirose, pp. 105–13.

36. For a discussion of the political resources at the disposal of big business, see Otake Hideo, *Gendai Nihon no seiji kenryoku, keizai kenryoku* (Political and economic power in contemporary Japan) (Tokyo: San'ichi Shobō, 1979), pp. 169–202.

37. For an issue-area or policy domain approach, see Dahl, and Theodore J. Lowi, "American Business, Public Policy Case-Studies, and Political Theory," *World Politics* 16 (July 1964): 677–715.

38. Otake Hideo's analysis of issues affecting the textiles and auto industries suggests that the policy-making processes for the two industries were handled in somewhat self-contained policy arenas; see Otake, *Gendai Nihon,* pp. 25–168.

39. Daniel I. Okimoto, "Domestic Political Configurations and Trade: The Grand Coalition of Interests," unpublished paper prepared for the Japan Political Economy Research Conference, East-West Center, Honolulu, Hawaii, Aug. 1984.

40. Mancur Olson, *The Rise and Decline of Nations* (New Haven, Conn.: Yale University Press, 1982), pp. 36–73.

41. Ibid., p. 51.

42. For a biting indictment of the politically motivated protection of the Japanese medical profession, see Watanabe Michio, "Han ishi yugu zeisei-ron" (Against the protection of doctors: A reform proposal), *Shokun,* March 1978.

43. The medical profession is used again as an example of economic inefficiency and politicization. See Itō Mitsuharu, "Gendai iryo keizai no mondaiten" (Problems in the economics of medical care), *Sekai,* June 1982, pp. 58–67.

44. See Haruhiro Fukui, *Party in Power: The Japanese Liberal-Democrats and Policymaking* (Berkeley: University of California Press, 1971); Nathaniel B. Thayer, *How the Conservatives Rule Japan* (Princeton, N.J.: Princeton University Press, 1969).

45. For a case study of a *kōenkai* in action, see Gerald L. Curtis, *Campaigning Japanese Style* (New York: Columbia University Press, 1971).

46. For a fascinating inside account of LDP factional politics, see Itō Masaya, *Jimintō sengoku-shi: Kenryoko no kenkyū* (A history of LDP internal warfare: Research into political power) (Tokyo: Asahi Sonorama, 1982), especially part 1, pp. 11–402.

47. Ishikawa, pp. 95–96.

48. Data for this analysis is derived from the district-by-district breakdown

of the 1982 Lower House elections, as provided in Miyakawa Takayoshi, ed., *Seiji handobukku: 1982* (1982 Handbook of Politics) (Tokyo: Seiji Kōhō Sentaa, 1982), pp. 233–76.

49. Itō Masaya, pp. 11–40.

50. The accepted rules of the game include the unusual precedent of having a party elder appoint a new party president when there is a succession crisis. See Mainichi Shinbun Seiji-bu, *Seihen* (Political change) (Tokyo: Mainichi Shinbunsha, 1975), pp. 9–29.

51. Interviews with seven LDP Diet members, Aug. 1983 and 1984.

52. Anthony Downs, *An Economic Theory of Democracy* (New York: Harper and Row, 1957).

53. Interviews with seven LDP Diet members, Aug. 1983 and 1984.

54. Joel D. Aberbach, Robert D. Putnam, and Bert A. Rockman, *Bureaucrats and Politicians in Western Democracies* (Cambridge, Mass.: Harvard University Press, 1981), pp. 9–23.

55. The idea of administrative reform had a precedent; a program by the same name had been undertaken in 1961. Shumpei Kumon "Japan Faces Its Future: The Political-Economics of Administrative Reform," *Journal of Japanese Studies* 10, no. 1 (Winter 1984): 145–46.

56. Interview with MITI official, Aug. 1984.

57. Gary R. Saxonhouse, "Comparative Advantage and Structural Adaptation," unpublished paper prepared for Japan Political Economy Research Conference, Honolulu, Hawaii, Aug. 5–11, 1984.

58. Haruhiro Fukui, "The Liberal Democratic Party Revisited: Continuity and Change in the Party's Structure and Performance," *Journal of Japanese Studies* 10, no. 2 (Summer 1984): 393.

59. Ibid.

60. Ōkubo Shōzō, *Hadaka no seikai* (The political world exposed) (Tokyo: Simul Press, 1975), pp. 7–10.

61. For a fascinating account of the formative years of the late 1940's and early 1950's recounted in a novel format, see Togawa Isamu, *Shōsetsu: Yoshida gakkō* (A novel: The yoshida school) (Tokyo: Ryūdō, 1971), especially pp. 85–129.

62. Tominomori Eiji, *Sengo hoshutō-shi* (A postwar history of the conservative party) (Tokyo: Nihon Hyōronsha, 1977), pp. 45–50.

63. Fukui, "Liberal Democratic Party," p. 393.

64. Ishikawa, pp. 30–32.

65. Tominomori, pp. 109–12 and 131–36.

66. I am grateful to Yuri Kondo, my research assistant, for painstakingly digging the data out of documents.

67. Ōkubo Shōzō, a veteran journalist on the political beat, has argued, in fact, that the most decisive difference between the LDP and the JSP has been the former's success at recruitment of bureaucrats and the latter's failure. Ōkubo, pp. 3–28.

68. Muramatsu Michio, *Sengo Nihon no kanryōsei* (The bureaucratic system in postwar Japan) (Tokyo: Nihon Hyōronsha, 1981), pp. 186–206.

69. Ishikawa, pp. 30–32.

70. For literary sarcasm that touches lightly on politics written in the form of poetry by a former high-ranking MITI official, see Amaya Naohiro, *Chōnin*

koku tedai no kurikotoba (Complaints of a merchant-state salesman) (Tokyo: Chūō Kōron Jigyō Shuppan, 1982).

71. Interviews with LDP politicians, MITI officials, and journalists, Apr. 1983 and Aug. 1984.

72. The information for this section is based on extensive interviews conducted over a two-year period with more than 30 LDP politicians (including members of IIPC and KIPC), MITI officials, members of industrial associations, and firsthand participants in and observers of the policy-making processes.

73. Kawaguchi Hiroyuki, "Zoku no rankingu" (The rankings of informal LDP support groups), *Kankai*, Nov. 1983, pp. 94–110.

74. Jun-ichi Kyogoku, *The Political Dynamics of Japan* (New York: Columbia University Press, 1987), pp. 117–32.

75. The most comprehensive presentation of the bureaucracy-dominant school can be found in Chalmers Johnson, "Japan: Who Governs? An Essay on Official Bureaucracy," *Journal of Japanese Studies* 2 (Autumn 1975): 1–28; interview data in support of the LDP-dominant school are found in Muramatsu Michio, *Sengo Nihon no kanryōsei*, pp. 137–68.

76. Masao Maruyama, *Thought and Behavior in Modern Japanese Politics*, ed. Ivan Morris (New York: Oxford University Press, 1963); Robert N. Bellah, *Tokugawa Religion: The Values of Pre-industrial Japan* (Glencoe, Ill.: Free Press, 1957); Chie Nakane, *Japanese Society* (Berkeley: University of California Press, 1970).

77. Bellah, *Tokugawa Religion*, pp. 11–57.

78. The Japanese state's role in maintaining a normative order is similar, in some respects, to the ceremonial functions performed by the state in nineteenth-century Bali. See Clifford Geertz, *Negara: The Theater State in Nineteenth-Century Bali* (Princeton, N.J.: Princeton University Press, 1981).

79. This discussion of the "organic state" draws heavily on Alfred Stepan's crisp and concise analysis in *The State and Society: Peru in Comparative Perspective* (Princeton, N.J.: Princeton University Press, 1978), pp. 26–45.

80. David Collier, *The New Authoritarianism in Latin America* (Princeton, N.J.: Princeton University Press, 1979), pp. 19–32.

81. Maruyama, p. 8.

82. See Barrington Moore's discussion of Japan's road to fascism, *Social Origins of Dictatorship and Democracy: Lord and Peasant in the Making of the Modern World* (Boston: Beacon Press, 1966).

Conclusions

1. See, e.g., Victoria Curzon Price, *Industrial Policies in the European Community* (London: St. Martin's Press, 1984).

2. Daniel I. Okimoto, "Political Power in Japan," in Shumpei Kumon and Henry Rosovsky, eds., *The Political Economy of Japan*, vol. 3, *Culture* (Stanford, Calif.: Stanford University Press, forthcoming).

3. Daniel I. Okimoto, "The Economics of National Defense," in D. I. Okimoto, ed., *Japan's Economy: Coping with Change in the International Environment* (Boulder, Colo.: Westview Press, 1982).

4. Daniel I. Okimoto, "The Japanese Challenge in High Technology," in

Ralph Landau and Nathan Rosenberg, eds., *The Positive Sum Strategy: Harnessing Technology for Economic Growth* (Washington, D.C.: National Academy Press, 1986), pp. 541–67.

5. Chie Nakane, *Japanese Society* (Berkeley: University of California Press, 1970), pp. 1–22.

6. Robert N. Bellah, *Tokugawa Religion: The Values of Pre-industrial Japan* (Glencoe, Ill.: Free Press, 1957).

7. Ronald P. Dore, *Taking Japan Seriously: A Confucian Perspective on Leading Economic Issues* (Stanford, Calif.: Stanford University Press, 1987).

Index

In this index an "f" after a number indicates a separate reference on the next page, and an "ff" indicates separate references on the next two pages. A continuous discussion over two or more pages is indicated by a span of page numbers, e.g., "57–59." *Passim* is used for a cluster of references in close but not consecutive sequence.

Library of Congress Cataloging-in-Publication Data

Okimoto, Daniel I., 1942–
   Between MITI and the market: Japanese industrial policy for
high technology / Daniel I. Okimoto.
   p.   cm.
   Includes index.
   ISBN 0-8047-1298-0 (alk. paper):
   ISBN 0-8047-1812-1 (pbk: alk. paper)
   1. High technology industries—Government policy—Japan.
2. Japan. Tsūshō Sangyōshō. I. Title.
HC465.H53035 1989   88-39837
338.4'762'000952—dc19   CIP